RANDOM MEDIA AND COMPOSITES

SIAM PROCEEDINGS SERIES LIST

Neustadt, L.W., Proceedings of the First International Congress on Programming and Control (1966)

Hull, T.E., Studies in Optimization (1970)

Day, R.H. & Robinson, S.M., Mathematical Topics in Economic Theory and Computation (1972)

Proschan, F. & Serfling, R.J., Reliability and Biometry: Statistical Analysis of Lifelength (1974)

Barlow, R.E., Reliability & Fault Tree Analysis: Theoretical & Applied Aspects of System Reliability & Safety Assessment (1975)

Fussell, J.B. & Burdick, G.R., Nuclear Systems Reliability Engineering and Risk Assessment (1977)

Duff, I.S. & Stewart, G.W., Sparse Matrix Proceedings 1978 (1979)

Holmes, P.J., New Approaches to Nonlinear Problems in Dynamics (1980)

Erisman, A.M., Neves, K.W. & Dwarakanath, M.H., Electric Power Problems: The Mathematical Challenge (1981)

Bednar, J.B., Redner, R., Robinson, E. & Weglein, A., Conference on Inverse Scattering: Theory and Application (1983)

Voigt, R.G., Gottlieb, D. & Hussaini, M. Yousuff, Spectral Methods for Partial Differential Equations (1984)

Chandra, Jagdish, Chaos in Nonlinear Dynamical Systems (1984)

Santosa, Fadil, Symes, William W., Pao, Yih-Hsing & Holland, Charles, Inverse Problems of Acoustic and Elastic Waves (1984)

Gross, Kenneth I., Mathematical Methods in Energy Research (1984)

Babuska, I., Chandra, J. & Flaherty, J., Adaptive Computational Methods for Partial Differential Equations (1984)

Boggs, Paul T., Byrd, Richard H. & Schnabel, Robert B., Numerical Optimization 1984 (1985)

Angrand, F., Dervieux, A., Desideri, J.A. & Glowinski, R., Numerical Methods for Euler Equations of Fluid Dynamics (1985)

Wouk, Arthur, New Computing Environments: Parallel, Vector and Systolic (1986)

Fitzgibbon, William E., Mathematical and Computational Methods in Seismic Exploration and Reservoir Modeling (1986)

Drew, Donald A. & Flaherty, Joseph E., Mathematics Applied to Fluid Mechanics and Stability: Proceedings of a Conference Dedicated to R.C. DiPrima (1986)

Heath, Michael T., Hypercube Multiprocessors 1986 (1986)

Papanicolaou, George, Advances in Multiphase Flow and Related Problems (1987)

Wouk, Arthur, New Computing Environments: Microcomputers in Large-Scale Computing (1987)

Chandra, Jagdish & Srivastav, Ram, Constitutive Models of Deformation (1987)

Heath, Michael T., Hypercube Multiprocessors 1987 (1987)

Glowinski, R., Golub, G.H., Meurant, G.A. & Periaux, J., First International Symposium on Domain Decomposition Methods for Partial Differential Equations (1988)

Salam, Fathi M.A. & Levi, Mark L., Dynamical Systems Approaches to Nonlinear Problems in Systems and Circuits (1988)

Datta, B., Johnson, C., Kaashoek, M., Plemmons, R. & Sontag, E., Linear Algebra in Signals, Systems and Control (1988)

Ringeisen, Richard D. & Roberts, Fred S., Applications of Discrete Mathematics (1988)

McKenna, James & Temam, Roger, ICIAM '87—Proceedings of the First International Conference on Industrial and Applied Mathematics (1988)

Rodrigue, Garry, Parallel Processing for Scientific Computing (1989)

Chan, Tony F., Meurant, Gerard, Periaux, Jacques & Widlund, Olof B., Domain Decomposition Methods (1989)

Caflisch, Russel E., Mathematical Aspects of Vortex Dynamics (1989)

Wouk, Arthur, Parallel Processing and Medium-Scale Multiprocessors (1989)

Flaherty, Joseph E., Paslow, Pamela J., Shephard, Mark S., & Vasilakis, John D., Adaptive Methods for Partial Differential Equations (1989)

Kohn, Robert V., and Milton, Graeme W., Random Media and Composites (1989)

RANDOM MEDIA AND COMPOSITES

Edited by Robert V. Kohn
and Graeme W. Milton
Courant Institute of Mathematical
Sciences, New York University

siam

Philadelphia

Society for Industrial and Applied Mathematics

RANDOM MEDIA AND COMPOSITES

Proceedings of the SIAM Workshop on Random Media and Composites, Leesburg, Virginia, December 7-10, 1988.

This workshop was supported by the United States Air Force Office of Scientific Research under grant no. AFOSR-86-0352.

Library of Congress Cataloging-in-Publication Data
SIAM Workshop on Random Media and Composites (1988 : Leesburg, Va.)
Random media and composites.

"Proceedings of the SIAM Workshop on Random Media and Composites, Leesburg, Virginia, December 7–10, 1988"–T. p. verso.
"Sponsored by the Air Force Office of Scientific Research"–Pref.
Includes bibliographical references.
1. Materials—Mathematical models—Congresses.
2. Composite materials—Mathematical models—Congresses.
I. Kohn, Robert V. II. Milton, Graeme W., 1956-
III. Society for Industrial and Applied Mathematics.
IV. United States. Air Force. Office of Scientific Research. V. Title.
TA401.3.S53 1988 620.1'18 89-26092

ISBN 0-89871-246-7

PREFACE

The Workshop on Random Media and Composites was held at the Xerox Training Center in Leesburg, Virginia in December, 1988. It attracted a diverse audience including specialists in theoretical and experimental physics, mechanical engineering, and geophysics, as well as applied mathematicians. The workshop covered a broad range of topics related to the behavior of disordered systems. Among the specific areas of focus were percolation and breakdown in random resistor networks, bounds and mean fields theories for the effective moduli of composites, seismic inverse problems, localization in random media, and superconductivity.

The workshop was sponsored by the Air Force Office of Scientific Research through the Center for Analysis of Heterogeneous and Nonlinear Media at the Courant Institute of Mathematical Sciences. The Center is funded by a University Research Initiative grant through the applied mathematics program of AFOSR, directed by Dr. Arje Nachman. The meeting was co-sponsored by SIAM, which handled the organizational aspects. The workshop program was determined by an organizing committee consisting of David Johnson (Schlumberger-Doll Research Laboratories), Robert Kohn (co-chairman, Courant Institute), Graeme Milton (co-chairman, Courant Institute), George Papanicolaou (Courant Institute), Pabitra Sen (Schlumberger-Doll Research Laboratories), Ping Sheng (Exxon Research and Engineering), Luc Tartar (Carnegie-Mellon University), and Benjamin White (Exxon Research and Engineering).

This volume contains research articles corresponding to 11 of the 29 talks presented. They are loosely organized into two categories: those primarily concerned with continuous systems, and those focussed on discrete systems. The volume also contains abstracts for most of the remaining talks. Both the research papers and the abstracts have extensive bibliographies, providing an invaluable guide to the recent scientific literature in this area.

The field of random media and composites is concerned with the modelling of many different types of materials, ranging from alloys, emulsions, and colloidal suspensions to resistor networks, polycrystalline aggregates, porous rocks, and fibrous materials. It is also concerned with many different types of physical phenomena, including electrical conduction, elastic response, and wave propagation. Its central goal is two-fold: one seeks to predict the properties of materials from partial knowledge of their fine scale structure; and one seeks to determine the essential features of the microstructure indirectly, from measurements of the overall physical properties.

This volume will be of interest to applied mathematicians and theoretically oriented physicists and engineers who are concerned with modelling the behavior of heterogeneous or random media. The research articles include many theoretical results not published elsewhere. The abstracts will help interested individuals explore other aspects of the field, particularly as it relates to experimental work.

The papers in this volume reflect the many fascinating problems that remain to be understood concerning random media and composite materials. We hope that their publication in this format will help to facilitate further progress in this technologically important field.

Robert V. Kohn
Graeme W. Milton

LIST OF CONTRIBUTORS

James G. Berryman, University of California, Lawrence Livermore National Laboratory, Livermore, California 94550

Safia Brahim-Otsmane, Laboratoire d'Analyse Numérique, Université Paris VI et CNRS, Tour 55-65, 75252 Paris Cedex 05, France

P. M. Duxbury, Department of Physics and Astronomy and Center for Fundamental Materials Research, Michigan State University, East Lansing, Michigan 48824-1116

Kenneth Golden, Department of Mathematics, Princeton University, Princeton, New Jersey 08544

Gilles A. Francfort, Laboratoire Central des Ponts et Chaussées, 75732 Paris Cedex 15, France

Y. S. Li, Department of Physics and Astronomy and Center for Fundamental Materials Research, Michigan State University, East Lansing, Michigan 48824-1116

C. J. Lobb, Division of Applied Sciences and Department of Physics, Harvard University, Cambridge, Massachusetts 02138

Francois Murat, Laboratoire d'Analyse Numérique, Université Paris VI et CNRS, Tour 55-56, 75252 Paris Cedex 05, France

L. Poladian, Sir Frank Packer Theoretical Physics Department, University of Sydney, NSW, 2006, Australia.

Sidney Redner, Center for Polymer Studies and Department of Physics, Boston University, Boston, Massachusetts 02215

Jacob Rubinstein, Department of Mathematics, Technion-Israel Institute of Technology, 32000 Haifa, Israel

William W. Symes, Department of Mathematical Sciences, Rice University, Houston, Texas 77005

Luc Tartar, Department of Mathematics, Carnegie Mellon University, Pittsburgh, Pennsylvania 15213-3890

Sal Torquato, Department of Mechanical and Aerospace Engineering, North Carolina State University, Raleigh, North Carolina 27695

J. R. Willis, School of Mathematical Sciences, University of Bath, Bath BA2 7AY, United Kingdom

CONTENTS

PART I

Continuous Media

CHAPTER 1

Estimating Effective Moduli of Composites Using Quantitative Image Analysis*

James G. Berryman†

Abstract. The statistical data required for quantitative analysis of the physical properties of composite materials may be obtained using digital image processing techniques. This approach allows accurate quantitative determination of volume fraction data, of interface area per unit volume, and of more sophisticated measures of the microgeometry of material mixtures including the n-point spatial correlation functions. The two- and three-point correlation functions can be measured quite accurately both for synthetic materials and for real materials such as porous sandstones. To extract the desired parameters of the microgeometry from the three-point correlation functions, methods of interpolating and integrating these lattice-based empirical values have been developed. Methods of finessing the theory to give estimates of physical properties when the available data are not adequate to permit use of the full theory have also been developed.

1. Introduction. Various methods of estimating effective properties of composite materials require geometrical or topological information contained in statistical correlation functions. If the three-point correlation functions are measured using digital image processing methods [1], the values are computed for a discrete set of admissible triangular arguments corresponding to triangles whose vertices are commensurate with a simple cubic lattice. To extract the desired information about microgeometry from the correlation functions, methods of interpolating and integrating between these lattice-based values have been developed [2].

---

* This work performed under the auspices of the U.S. Department of Energy by the Lawrence Livermore National Laboratory under contract No. W-7405-ENG-48 and supported specifically by the DOE Office of Basic Energy Sciences.

† University of California, Lawrence Livermore National Laboratory, P.O. Box 808, L-156, Livermore, CA 94550.

Whereas the variational bounds derived by Hashin and Shtrikman [3] for elastic constants depend only on the volume fractions and the moduli of the constituents of the composite, nearly all of the improved bounds on physical properties require additional information about the microstructure of the composite. Perhaps the most common method of quantifying this microstructural information comes in the form of the n-point spatial correlation functions. These functions were introduced into this problem by Brown [4] in his early work on obtaining estimates of effective constants in series form. Debye *et al.* [5] used the angular distribution of scattered x-rays to measure the two-point spatial correlation function and estimate the specific (internal) surface area of porous materials. Prager [6] used these same functions in his early formulation of variational bounds for the fluid permeability of porous media. Beran [7] also needed them in his variational approach for finding bounds on the dielectric constant. Theoretical results that require the information contained in the correlation functions have subsequently been obtained by many authors over a period of many years [8-16], and we may expect more applications will be found in the future. It is therefore clear that methods of measuring these functions are needed in order to make these theoretical results accessible to experimentalists and engineers interested in designing new materials. Corson [13] was the first to try to measure the three-point correlation function, but — with the advent of modern image processing techniques and faster computers — better techniques are available to us now [1,17,18] than were available to him then.

In this paper we will define the correlation functions in Section 2 and present some of their more important general properties. Practical approximation methods are also discussed. In Section 3 we discuss the parameters of microgeometry that are commonly encountered in studies of electrical, thermal, and elastic composites. To provide some new insight into the structure of the correlation functions, we present a new method of visualizing the three-point correlation functions in Section 4.

2. Spatial correlation functions. A discussion of the significance of the spatial correlation functions has been presented by Torquato and Stell [14] and a detailed analysis of methods for obtaining these functions experimentally using image processing techniques was presented in [1]. We will not repeat these detailed discussions here, but it is still necessary to define the correlation functions and mention their relevant properties. The analysis presented is limited to two-phase composite media.

Let $p(\vec{x})$ be the value of some property of a random composite material (e.g., electrical or thermal conductivity, dielectric constant, bulk or shear modulus, etc.) which assumes one of two values p_0 or p_1 depending on whether $\vec{x}$ is located in a grain of material 0 or material 1. Define the indicator or characteristic function

$$f(\vec{x}) = \frac{p(\vec{x}) - p_0}{p_1 - p_0}. \tag{1}$$

Then $f(\vec{x}) = 0$ in material 0 and $f(\vec{x}) = 1$ in material 1. For example, in a porous medium we may arbitrarily label all solid regions as material 0 and all void regions as material 1. Since complete knowledge of the stochastic variable f is seldom available, our interest in the characteristic function is generally limited to a few

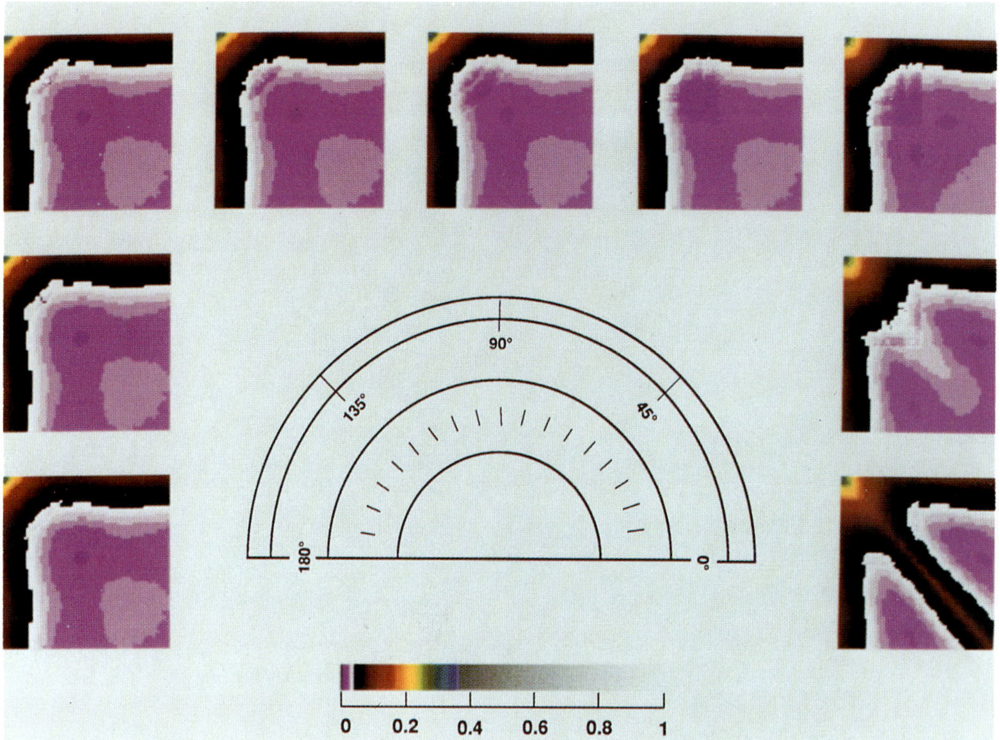

Figure 2. Display of the three-point correlation function $S_3(r, s, \cos\theta)$ as the angle θ between the sides r, s of the triangular argument varies from $0°$ to $180°$ in steps of $22.5°$. In each frame, the origin for the (r, s)-plane is in the upper left corner.

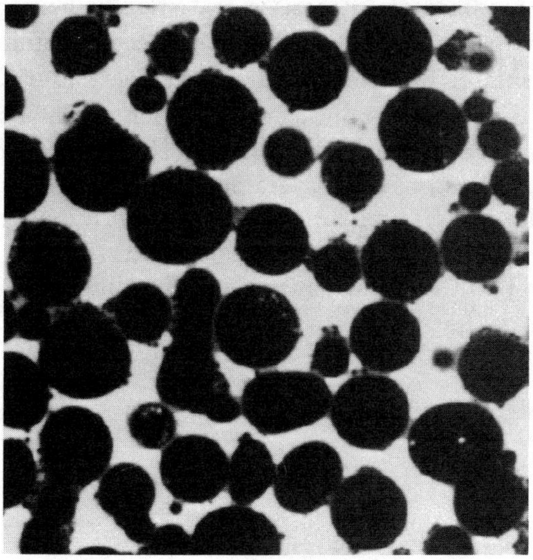

Figure 1. Pseudo-gray scale image of a cross section of a glass bead sample obtained by digitizing a raw scanning electron microscope image. Dark areas are solid and light areas are epoxy that has been injected into the pore space. This image must be registered (ones or zeroes must be assigned to every pixel) before the spatial correlation functions are computed. To establish the scale of this image, the apparent diameter of the largest particles is approximately 100μ.

of its statistical properties. If chosen properly, these quantities are often sufficient to provide the data needed for variational bounds on the macroscopic average of the property being studied [1]. Figure 1 shows an example of a scanning electron microscope image of a cross section of a real material (a glass bead sample) prior to registration.

The first three spatial correlation functions are defined by

$$\hat{S}_1 = < f(\vec{x}) > = \phi, \tag{2}$$

$$\hat{S}_2(\vec{r}_1, \vec{r}_2) = < f(\vec{x} + \vec{r}_1) f(\vec{x} + \vec{r}_2) >, \tag{3}$$

and

$$\hat{S}_3(\vec{r}_1, \vec{r}_2, \vec{r}_3) = < f(\vec{x} + \vec{r}_1) f(\vec{x} + \vec{r}_2) f(\vec{x} + \vec{r}_3) > . \tag{4}$$

The brackets $< \cdot >$ indicate a volume average over the spatial coordinate $\vec{x}$. The volume fraction of constituent 1 is given by ϕ. We will refer to the three correlation functions defined this way as the one-, two-, and three-point correlation functions, respectively. Since two points lie along a line and three points lie in a plane, the two-point and three-point correlations (as well as the one-point correlation) may be measured by processing digital images of material cross sections. In the present paper, we will stress the three-point correlation functions. In general, we assume that the composite medium of interest is statistically homogeneous so that on average only the differences in the coordinate values are significant (translational invariance). Furthermore, we often assume that the material is at least locally isotropic so that the averages do not depend on orientation of the arguments. (An example of local isotropy in conjunction with global anisotropy is a transversely isotropic material composed of thin layers of isotropic materials. Such a material can be treated if the correlation functions are determined by examining cross sections taken perpendicular to the axis of symmetry.) With these assumptions, we find that the two-point correlation function simplifies to

$$\hat{S}_2(\vec{r}_1,\vec{r}_2) = \tilde{S}_2(\vec{r}_{12}) = S_2(r_{12}), \tag{5}$$

while the three-point correlation function satisfies

$$\hat{S}_3(\vec{r}_1,\vec{r}_2,\vec{r}_3) = \tilde{S}_3(\vec{r}_{12},\vec{r}_{13}) = S_3(r_{12},r_{13},\mu_{12,13}), \tag{6}$$

where

$$\vec{r}_{ij} = \vec{r}_j - \vec{r}_i, \qquad r_{ij} = \mid \vec{r}_{ij} \mid,$$

and

$$\mu_{ij,ik} = cos\,\theta = \vec{r}_{ij}\cdot\vec{r}_{ik}/r_{ij}r_{ik}.$$

The two-point correlation $S_2(r)$ is the probability that two points a distance $r = r_{12}$ apart are both in material 1. The three-point correlation function $S_3(r,s,\mu)$ is the probability that all three vertices of the triangle determined by (r,s,μ) lie in material 1.

In a random material possessing no long range order, we have

$$\lim_{s\,fixed, r\to\infty} S_3(r,s,\mu) = \phi S_2(s) \tag{7}$$

and

$$\lim_{r,s\to\infty} S_3(r,s,\mu) = \phi^3, \tag{8}$$

assuming that $\mu \neq 1$ or that the difference $r - s$ is not fixed if $\mu = 1$. The most important property of S_3 for statistically homogeneous and isotropic media is the fact that the value depends only on the size and shape of the triangle formed by its arguments, not on the orientation of that triangle. Thus,

$$S_3(r_{12},r_{13},\mu_{12,13}) = S_3(r_{21},r_{23},\mu_{21,23}) = S_3(r_{31},r_{32},\mu_{31,32}). \tag{9}$$

Furthermore, the order of the first two arguments may be freely interchanged. It is important to account for the symmetries (9) when designing a scheme to calculate, sort, and store the values of S_3, otherwise the stored values could be as much as sixfold degenerate. Such a scheme has been described in [2].

In general, these spatial correlation functions must be determined empirically. These one-, two-, and three-point correlation functions may be found using digital image processing techniques described previously [1]. Accurate values of S_3 are not difficult to obtain for small triangles whose vertices fit on the discrete lattice composed of the pixels in a digitized image (we call such triangles "lattice-commensurate"). However, there are several practical reasons why it is virtually impossible to obtain accurate values of the three-point correlation functions for large vertex separations: (*i*) The number of lattice-commensurate triangles of large size increases like the cube of the length of the longest side of the triangle (in pixel units), so storage of computed values of S_3 rapidly becomes difficult. (*ii*) As the triangle becomes large, fewer and fewer of them will fit on an image of fixed size, so the size of the statistical sample of S_3 for large arguments rapidly decreases and accuracy therefore suffers. Since the integrals used to define the parameters of the microgeometry are over triangles of all sizes, these practical limitations to our knowledge of S_3 must be accounted for in any image analysis scheme we devise.

Another significant practical issue arises due to the structure of the correlation function itself. Although the asymptotic value of $S_3(r,s,\mu)$ for large triangles is ϕ^3 as stated in (8), there are three exceptions and two are important: (*i*) If either $r=0$ or $s=0$, then the triangle collapses to a straight line and the value of the angle cosine μ is immaterial. In this situation,

$$S_3(0,s,\mu)=S_2(s) \qquad \text{or} \qquad S_3(r,0,\mu)=S_2(r). \tag{10}$$

However, since these results are independent of μ, we will see that (10) causes no problems in the integrals for the parameters of microgeometry. (*ii*) On the other hand, if $r=s$, then with $\mu=1$ the triangle again collapses to a straight line and we find

$$S_3(r,r,1)=S_2(r). \tag{11}$$

For large r, we find that S_3 approaches ϕ^2 instead of ϕ^3 for either (10) or (11), but (11) is more troublesome because it does depend on angle. (*iii*) Similarly, if $r\simeq s$ and r is large, then for any value of μ it is easy to see that

$$S_3(r,s,\mu)\simeq\phi S_2(t), \tag{12}$$

where $t=(r^2+s^2-2rs\mu)^{\frac{1}{2}}$ is the length of the third side of the triangle in argument of S_3. Eq. (12) follows from (4) by noting that if both r and s are large while t is small, then one of the three vertices of the triangle is far from the other two and therefore only weakly correlated with them. Thus, the distant point may be removed from the average in (4) and its contribution included through the factor ϕ in (12), *c.f.*, Eq. (7). Note that (12) is an approximation while (11) is exact. As $t\to 0$, we see that (12) smoothly approaches (11) only for such large values of r that $S_2(r)=\phi^2$.

To treat the special regions of S_3 noted in the last paragraph and to provide an approximate solution to the problem of finding accurate values of S_3 for large triangular arguments, we introduce the following approximation when at least one of the two sides r,s is large:

$$S_3(r,s,\mu)\simeq S_2(x)S_2(y)/\phi, \tag{13}$$

where x and y are the two shortest sides of the triangle whose three sides are r,s,t. When $r=0$, $x=r=0$ and $y=s$ so (13) gives the exact result (10). When $r=s$ and $\mu=1$, $x=t=0$ and $y=r$ so (13) gives the exact result (11). When $r\simeq s$ and $t<r$, $x=t$ and $y=$ the smaller of r and s (say r), then (13) gives

$$S_3(r,s,\mu)\simeq S_2(t)S_2(r)/\phi, \tag{14}$$

which reduces to (12) when r is large, as we have assumed it is.

Note that when $x = y = 0$, (13) gives the exact result $S_3(0,0,\mu) = \phi$ and, when x and y are both large, (13) gives the correct asymptotic limit $S_3 \to \phi^3$. Thus, although (13) is an approximation, it does reproduce all the special limits we have discussed. Furthermore, the use of (13) for large arguments solves the problem of obtaining accurate values of S_3 for large arguments and also eliminates the storage problem, since S_2 is much easier to compute and store than S_3. Finally, note that the approximation (13) is superior to other approximations that have been used [19] such as $S_3(r,s,\mu) = S_2(r)S_2(s)/\phi$ which has no angular dependence (and therefore would produce nothing if used in the integrals) or such as $S_3(r,s,\mu) = S_2(r)(S_2(s) + S_2(t))/2\phi$ which does not satisfy all of the limiting conditions.

We will show an example later that illustrates the accuracy of the approximation (13) for large arguments.

3. Parameters of microgeometry. Milton [15,16] has introduced two parameters depending on the microgeometry of a composite through the three-point correlation function S_3. These two geometric parameters of interest for studies of electrical or thermal conductivities and elastic constants are

$$\zeta_1 = \lim_{\Delta\to 0}\lim_{\Delta'\to\infty} \frac{9}{2\phi_0\phi_1}\int_\Delta^{\Delta'} dr \int_\Delta^{\Delta'} ds \int_{-1}^{+1} d\mu \frac{S_3(r,s,\mu)}{rs} P_2(\mu) \tag{15}$$

and

$$\eta_1 = \frac{5\zeta_1}{21} + \lim_{\Delta\to 0}\lim_{\Delta'\to\infty} \frac{150}{7\phi_0\phi_1}\int_\Delta^{\Delta'} dr \int_\Delta^{\Delta'} ds \int_{-1}^{+1} d\mu \frac{S_3(r,s,\mu)}{rs} P_4(\mu), \tag{16}$$

where $\phi_1 = \phi$, $\phi_0 = 1-\phi_1$, and $P_2(\mu)$ and $P_4(\mu)$ are the Legendre polynomials of order 2 and 4 given respectively by

$$P_2(\mu) = \frac{1}{2}(3\mu^2 - 1)$$

and

$$P_4(\mu) = \frac{1}{8}(35\mu^4 - 30\mu^2 + 3).$$

The complements of ζ_1 and η_1 are $\zeta_0 = 1-\zeta_1$ and $\eta_0 = 1-\eta_1$, respectively. All four of these constants lie in the range [0,1].

The integrands of both integrals (15) and (16) are singular for small values of r, s and poorly behaved for large values of r, s. The conditional convergence of the integrals requires that the integration over μ be performed prior to the integrals over r and s. The singularity at the origin may be handled easily by standard methods [20]. The exact relations (10) guarantee that triangles with at least one side of vanishing length have no angular dependence and therefore no contribution to these integrals.

Recalling the important fact (12), we can improve the rate of convergence at large values of the independent variables r and s by adding and subtracting a term that can be integrated analytically [16] giving

$$\zeta_1 = \phi_1 + \lim_{\Delta\to 0}\lim_{\Delta'\to\infty} \frac{9}{2\phi_0\phi_1}\int_\Delta^{\Delta'} dr \int_\Delta^{\Delta'} ds \int_{-1}^{+1} d\mu \frac{S_3(r,s,\mu) - \phi_1 S_2(t)}{rs} P_2(\mu) \tag{17}$$

and

$$\eta_1 = (5\zeta_1 + 16\phi_1)/21 + \lim_{\Delta\to 0} \lim_{\Delta'\to\infty} \frac{150}{7\phi_0\phi_1} \int_\Delta^{\Delta'} dr \int_\Delta^{\Delta'} ds \int_{-1}^{+1} d\mu \frac{S_3(r,s,\mu) - \phi_1 S_2(t)}{rs} P_4(\mu), \quad (18)$$

where $t = (r^2 + s^2 - 2rs\mu)^{\frac{1}{2}}$. The integrals of S_3 may be evaluated using numerical methods described in detail in references [2] and [20], or they may be evaluated more simply using methods we will describe in Section 4.

A physical interpretation for the parameters ζ_1 and η_1 is lacking at the present time. However, it is known that for the relevant applications (effective electrical or thermal conductivity, dielectric constant, elastic constants, etc.) the effective constants are themselves scale invariant and, therefore, any pertinent parameters of the microgeometry should also be scale invariant. Eqs. (15) and (16) clearly satisfy the scale invariance criterion, and it is hard to construct any other scale invariant scalars depending on S_3 or integrals of S_3 (at least this author has not been able to find any). The scale invariance argument also shows why the Hashin-Shtrikman bounds depend only on volume fraction information even though S_2 appears at intermediate stages of the derivation [9].

One significant practical observation about the results for ζ and η concerns the choice of image magnification. The digital image used to produce the empirical values of S_3 may or may not contain the level of detail required for accurate computation of these integrals. We have found [2] that our integration scheme produces good results for ζ when the effective radius R of the penetrable spheres satisfies $R/H \simeq 12$ where H is the pixel width. For η the best results were obtained when the effective radius of the penetrable spheres satisfied $R/H \simeq 25$. Thus, the image magnification required for the two integrals differs by about a factor of 2. This result is very reasonable when we compare the Legendre polynomials P_2 and P_4 appearing in the integrands. Since P_4 has twice as many zeroes as P_2, it is clear that, whatever resolution is required to obtain accurate values of ζ, about twice that resolution will be needed to obtain comparable accuracy for η. The observed behavior conforms to this expectation.

Although we expect this factor of 2 difference in optimum image magnification to be maintained for real materials, we do not expect that these particular values of (particle radius/pixel size) will be appropriate for all materials. The spherical particle shape and the lack of both cracks and surface roughness make the penetrable sphere model somewhat unrealistic. If the magnification of an image must be increased substantially beyond the values quoted in the preceding paragraph, then it is likely that the number of pixels in the image will also have to increase beyond the value of 512×512 assumed here. With the largest triangles having side $l \leq 64$, the total number of S_3 values computed from an image is 47377. Increasing the largest triangle size by a factor of 2 means increasing the total number of S_3 values to 364065 [2]; such an increase entails storing more computed values of S_3 than the total number of pixels in the original image. If such an increase in the data base is really needed for accurate determination of the integrals of interest, it might be preferable to use the digitized image itself as the data base for the larger triangles and compute the values of S_3 for these triangles as needed during the integration. Another approach would be to increase the number of pixels per image to 1024×1024 or 2048×2048 directly or by constructing a mosaic. The interpolation technique developed in [2] could be used with any of these approaches to the integration problem, and is used with still another approach in the next section.

4. Visualization of three-point correlations. Because the three-point correlation function $S_3(r, s, \cos\theta)$ is a function of triangle shape and size and therefore of three arguments, it is somewhat difficult to display the function. If we fix the angle θ between the sides r, s of the triangle, then it is possible to display the resulting function of two variables. The display can take the form of a contour plot in the (r, s)-plane showing lines of constant $S_3(r, s, \mu)$; the display could be performed using three-dimensional plotting routines that actually show the surface of $S_3(r, s, \mu)$ above the (r, s)-plane; or the display can be presented in color as we will do here (see Figure 2). The color display has the advantage of a contour plot in that all of the function can be shown simultaneously (three-dimensional surface plots always suffer from the fact that part of the surface is hidden behind the peaks of the high values closest to the viewer). Furthermore, it has the advantage that quantitative information about the values of the function are coded into the color mapping.

Figure 2 shows nine views of the three-point correlation function computed from the cross section of the glass bead sample shown in Figure 1. The nine views were chosen for equally spaced angles from $\theta = 0°$ to $\theta = 180°$. The superimposed protractor shows clearly what the angle is for each frame displayed. The color bar at the bottom shows the mapping between color and the numerical value of the three-point correlation function. Since the porosity (volume fraction of void space) for the glass bead sample was around 30%, the spectrum of colors has been compressed in this example so the interesting features are most easily seen.

There were 512×512 pixels in Figure 1. The two-point correlation function was computed to a lag of 256 pixel widths. The full three-point correlation function was computed only for triangles with largest side ≤ 64 pixel widths, and the display in each frame of Figure 2 shows the composite three-point correlation function for triangles with largest side ≤ 128 pixel widths. The values for S_3 in the upper left corner of each frame were computed using the interpolation scheme of [2]. Note that the region where this interpolation scheme is used depends on the angle: For $\theta = 0°$, this region is a 64×64 pixel square. For $\theta = 90°$, this region is one-quarter of a circle with 64 pixel radius. For $\theta = 180°$, this region is an isosceles right triangle with adjacent sides 64 pixels in length. The remaining values of S_3 were obtained using the approximation (13). The accuracy of (13) for large arguments is demonstrated in Figure 2 by the high degree of continuity observed between the regions calculated using the different methods.

The main features of the three-point correlation function are easily deduced from this display. All nine frames show the characteristic behavior along the top and left hand sides that the three-point correlation function is independent of angle if either $r = 0$ or $s = 0$ — the reason being that if one side of the triangle vanishes then the triangle reduces to a line and the three-point correlation function is just the two-point correlation function for the length of that line. The single frame that is most distinct from all the others is for $\theta = 0°$. When the angle between r and s vanishes and $r = s$, then again the value of the three-point correlation function is given by the two-point correlation function $S_2(r)$. This behavior dominates the frame in the lower right of Figure 2. As the angle increases monotonically from zero, we see that the correlation function becomes nearly independent of angle for $\theta > 90°$, but the significantly higher values of the correlation function at small angles when $r = s$ is still very apparent for $\theta = 22.5°$.

Once we have the capability of displaying the three-point correlation function for different choices of angle as in Figure 2, it is easy to make the next step of computing the parameters of the microgeometry ζ and η. To do so, we must first do the angular integration by multiplying the values of S_3 by the appropriate values of the Legendre polynomials for each angle and then combining them in quadrature. This step produces the integrals $\int d\mu S_3(r,s,\mu)P_2(\mu)$ and $\int d\mu S_3(r,s,\mu)P_4(\mu)$. The remaining integrals over the square domain in the (r,s)-plane are straightforward. The accuracy of the resulting values of ζ and η will depend strongly on the choice of image magnification used. In particular, the examples chosen for display in Figure 2 were chosen to illustrate the characteristics of the correlation function, but higher resolution is required to obtain accurate values of the parameters of microgeometry if this integration scheme is to be used. Other integration schemes have been used to produce more accurate results [2, 19, 20].

5. Discussion. We have shown that it is possible to obtain useful measured values of three-point correlation functions for real materials, to interpolate and integrate those values to find the parameters of the composite microgeometry, and to use new visualization techniques to gain insight into the structure of the correlation functions themselves. If our information is incomplete — so that accurate values of both ζ and η cannot be obtained from image analysis techniques, other methods of attack may be required. It has been shown elsewhere [21] that much can be learned about effective properties when other information is available. For example, the existence of the bounds on elastic constants using the microgeometry parameters is enough to show that some improvements can be made to the Hashin-Shtrikman bounds.

Acknowledgments. I thank Marco Avellaneda and Graeme W. Milton for their insight concerning the importance of scale invariance for the analysis of some physical properties. I thank Stephen C. Blair for his continued efforts to make quantitative image analysis a practical tool for estimating physical properties of rocks.

REFERENCES

[1] J. G. BERRYMAN, *Measurement of spatial correlation functions using image processing techniques*, J. Appl. Phys., **57** (1985), pp. 2374-2384.

[2] J. G. BERRYMAN, *Interpolating and integrating three-point correlation functions on a lattice*, J. Comput. Phys., **75** (1988), pp. 86-102.

[3] Z. HASHIN and S. SHTRIKMAN, *A variational approach to the theory of the elastic behaviour of multiphase materials*, J. Mech. Phys. Solids, **11** (1963), pp. 127-140.

[4] W. F. BROWN, JR., *Solid mixture permittivites*, J. Chem. Phys., **23** (1955), pp. 1514-1517.

[5] P. DEBYE, H. R. ANDERSON, JR., and H. BRUMBERGER, *Scattering by an inhomogeneous solid. II. The correlation function and its application*, J. Appl. Phys., **28** (1957), pp. 679-683.

[6] S. PRAGER, *Viscous flow through porous media*, Phys. Fluids, **4** (1961), pp. 1477-1482.

[7] M. J. BERAN, *Use of the variational approach to determine bounds for the effective permittivity in random media*, Nuovo Cimento, **38** (1965), pp. 771-782.

[8] M. J. BERAN and J. MOLYNEUX, *Use of classical variational principles to determine bounds for the effective bulk modulus in heterogenous media*, Quart. Appl. Math., **24** (1965), pp. 107-118.

[9] W. F. BROWN, JR., *Dielectric constants, permeabilities, and conductivities of random media*, Trans. Soc. Rheol., **9** (1965), pp. 357-380.

[10] M. J. BERAN, *Statistical Continuum Theories*, Interscience, New York, 1968.

[11] J. J. McCOY, *On the displacement field in an elastic medium with random variations of material properties*, in Recent Advances in Engineering Sciences, Vol. 5, A. C. Eringen, ed., Gordon and Breach, New York, 1970, pp. 235-254.

[12] M. J. BERAN and N. R. SILNUTZER, *Effective electrical, thermal and magnetic properties of fiber reinforced materials,* J. Compos. Mat., **5** (1971), pp. 246-249.

[13] P. B. CORSON, *Correlation functions for predicting properties of heterogeneous materials*, J. Appl. Phys., **45** (1974), pp. 3159-3182.

[14] S. TORQUATO and G. STELL, *Microstructure of two-phase random media. I. The n-point probability functions*, J. Chem. Phys., **77** (1982), pp. 2071-2077.

[15] G. W. MILTON, *Bounds on the electromagnetic, elastic, and other properties of two-component composites*, Phys. Rev. Lett., **46** (1981), pp. 542-545.

[16] G. W. MILTON, *Bounds on elastic and transport properties of two-component composites*, J. Mech. Phys. Solids, **30** (1982), pp. 177-191.

[17] J. G. BERRYMAN and S. C. BLAIR, *Use of digital image analysis to estimate fluid permeability of porous materials: Application of two-point correlation functions*, J. Appl. Phys., **60** (1986), pp. 1930-1938.

[18] J. G. BERRYMAN and S. C. BLAIR, *Kozeny-Carman relations and image processing methods for estimating Darcy's constant*, J. Appl. Phys., **62** (1987), pp. 2221-2228.

[19] S. TORQUATO and G. STELL, *Microstructure of two-phase random media. III. The n-point matrix probability functions for fully penetrable spheres*, J. Chem. Phys., **79** (1983), pp. 1505-1510.

[20] J. G. BERRYMAN, *Variational bounds on elastic constants for the penetrable sphere model*, J. Phys. D: Appl. Phys., **18** (1985), pp. 585-597.

[21] J. G. BERRYMAN and G. W. MILTON, *Microgeometry of random composites and porous media*, J. Phys. D: Appl. Phys., **21** (1988), pp. 87-94.

CHAPTER 2

Homogenization in Thermoelasticity

Safia Brahim-Otsmane*
Gilles A. Francfort†
François Murat*

ABSTRACT.

This paper is concerned with the homogenization of a system of partial differential equations which describes the thermomechanical behaviour of an heterogeneous medium. The unknowns are the linearized displacement u^ε and temperature increment τ^ε, where ε denotes the typical size of the heterogeneities.

After recalling results of existence and uniqueness, we pass to the limit as ε tends to zero. The limit (u, τ) of $(u^\varepsilon, \tau^\varepsilon)$ happens to be the solution of a system of the same type. The specific heat coefficient is however modified in an unexpected manner. More surprizing is the fact that the initial value for the temperature is also altered.

The changes are analyzed through a study of correctors in the last part of the paper.

* Laboratoire d'Analyse Numérique, Université Paris VI et CNRS, Tour 55-65, 75252 PARIS CEDEX 05, France.

† Laboratoire Central des Ponts et Chaussées, 58 Boulevard Lefebvre, 75732 PARIS CEDEX 15, France.

1. SETTING OF THE PROBLEM.

The following system is investigated:

$$\rho^\varepsilon \frac{\partial^2 u^\varepsilon}{\partial t^2} - div(A^\varepsilon\, grad\, u^\varepsilon - \gamma^\varepsilon \tau^\varepsilon) = f \quad \text{in} \quad \Omega \times (0,T) \tag{1.1}$$

$$\beta^\varepsilon \frac{\partial \tau^\varepsilon}{\partial t} - div(K^\varepsilon\, grad\, \tau^\varepsilon) + \gamma^\varepsilon\, grad \frac{\partial u^\varepsilon}{\partial t} = g \quad \text{in} \quad \Omega \times (0,T) \tag{1.2}$$

$$u^\varepsilon(x,t) = 0 \quad \text{on} \quad \partial\Omega \times (0,T) \tag{1.3}$$

$$\tau^\varepsilon(x,t) = 0 \quad \text{on} \quad \partial\Omega \times (0,T) \tag{1.4}$$

$$u^\varepsilon(x,0) = a^\varepsilon(x) \quad \text{in} \quad \Omega \tag{1.5}$$

$$\frac{\partial u^\varepsilon}{\partial t}(x,0) = b^\varepsilon(x) \quad \text{in} \quad \Omega \tag{1.6}$$

$$\tau^\varepsilon(x,0) = c^\varepsilon(x) \quad \text{in} \quad \Omega. \tag{1.7}$$

This system is a scalar version of the usually accepted equation modeling linear thermoelasticity (see [13], [4]) and consists, besides the boundary and initial conditions (1.3)-(1.7) of the coupling of a hyperbolic equation with a parabolic one. The displacement field is thus scalar valued which reduces the cumbersome notation of elasticity. Such a simplification does not alter the mathematical treatment of the problem (see [5] for the case where u^ε is a vector valued displacement field).

In this system the unknowns are the linearized displacement $u^\varepsilon(x,t)$ and temperature increment $\tau^\varepsilon(x,t)$. All other quantities are data of the problem : Ω is a bounded open subset of $\mathbb{R}^N$ with boundary $\partial\Omega$ (not necessarily smooth) ; T is a positive real number. The coefficients ρ^ε , β^ε , A^ε , K^ε and γ^ε are assumed to depend only on $x \in \Omega$, to be measurable and bounded and to satisfy

$$\begin{cases} \rho^\varepsilon \in L^\infty(\Omega) \quad , \quad \beta^\varepsilon \in L^\infty(\Omega) \\ \lambda_1 \le \rho^\varepsilon(x) \le \lambda_2 \quad , \quad \lambda_1 \le \beta^\varepsilon(x) \le \lambda_2 \quad \text{a.e.} \quad x \in \Omega \end{cases} \tag{1.8}$$

(1.9)
$$\begin{cases} A^\varepsilon \in (L^\infty(\Omega))^{N^2} \quad , \quad K^\varepsilon \in (L^\infty(\Omega))^{N^2} \\ A^\varepsilon = {}^tA^\varepsilon \quad , \quad K^\varepsilon = {}^tK^\varepsilon \\ \lambda_1|\zeta|^2 \le A^\varepsilon(x)\zeta\zeta \le \lambda_2|\zeta|^2 \, , \, \lambda_1|\zeta|^2 \le K^\varepsilon(x)\zeta\zeta \le \lambda_2|\zeta|^2 \, , \text{ a.e. } x \in \Omega \, , \forall\zeta \in \mathbb{R}^N \end{cases}$$

$$\begin{cases} \gamma^\varepsilon \in (L^\infty(\Omega))^N \\ |\gamma^\varepsilon(x)| \le \lambda_2 \quad \text{a.e.} \quad x \in \Omega \end{cases} \tag{1.10}$$

where $0 < \lambda_1 < \lambda_2$ are given real numbers. See [13], [4] for the mechanical significance of these quantities. The loading f and heat source g are assumed to satisfy

$$f \in L^2(0,T;L^2(\Omega)) \quad , \quad g \in L^2(0,T;H^{-1}(\Omega)) \tag{1.11}$$

(more general right hand sides could be considered), and the initial data a^ε , b^ε , c^ε are such that

$$\left\{\begin{array}{lll} a^\varepsilon \in H_0^1(\Omega) & , & \|a^\varepsilon\|_{H_0^1(\Omega)} \leq \lambda_3 \\ b^\varepsilon \in L^2(\Omega) & , & \|b^\varepsilon\|_{L^2(\Omega)} \leq \lambda_3 \\ c^\varepsilon \in L^2(\Omega) & , & \|c^\varepsilon\|_{L^2(\Omega)} \leq \lambda_3 \end{array}\right. \tag{1.12}$$

for some positive real number λ_3.

Finally ε denotes a parameter which takes its values in a sequence and tends to 0. This parameter describes the typical size of the heterogeneities of the medium.

In this paper we revisit results from the theses of Gilles A. Francfort and Safia Brahim-Otsmane (see [2], [4], [5], [6]). The three last references are concerned with a periodic setting where ρ^ε , β^ε , A^ε , K^ε and γ^ε are given by formulas of the type $\rho^\varepsilon(x) = \rho(x/\varepsilon)$ with ρ periodic. A few of the results presented in the present paper are new, for example the fact that σ belongs to $L^\infty(\Omega)$ and the study of the correctors in spaces of time-continuous functions.

The plan is as follows : Section 2 briefly recalls existence and uniqueness results. The question of the sense of the initial conditions is emphasized because it plays an important role in the change of the initial condition for the temperature. Section 3 defines the quantities appearing in the homogenized system and states the homogenization result for (1.1)-(1.7). With respect to (1.1)-(1.7) the homogenized system is of the same type as the original one (which is in itself an important feature). It exhibits however two essential differences : on the one hand, the heat conduction coefficient is modified ; on the other hand, the initial condition for the temperature is completely changed. Section 4 gives the proofs of the homogenization result, essentially by a reduction process leading to an elliptic setting, and by a carefull study of the sense of the initial conditions. In Section 5 these changes are investigated through a study of the correctors : the solution $(u^\varepsilon , \tau^\varepsilon)$ is split into two parts $(\tilde{u}^\varepsilon , \tilde{\tau}^\varepsilon)$ and $(v^\varepsilon , \theta^\varepsilon)$. The first part $(\tilde{u}^\varepsilon , \tilde{\tau}^\varepsilon)$ converges to the same limit as $(u^\varepsilon , \tau^\varepsilon)$ but with a "correct energy" which gives rise to a corrector result. There is no change in the initial conditions for $(\tilde{u}^\varepsilon , \tilde{\tau}^\varepsilon)$. The second part $(v^\varepsilon , \theta^\varepsilon)$ converges weakly to $(0,0)$ and dissipates the part of the energy which complements the "correct energy" of $(\tilde{u}^\varepsilon , \tilde{\tau}^\varepsilon)$. Proofs of the corrector results are presented in Section 6. They use the same ideas as for the elliptic setting and for the heat and wave equations (see [3]), but many details are more delicate to handle.

2. EXISTENCE AND UNIQUENESS RESULTS.

The main feature of system (1.1)-(1.7) resides in the following estimate formally obtained by multiplication of the equations (1.1), (1.2) by $\partial u^\varepsilon/\partial t$ and θ^ε respectively, integration over Ω and integration by parts using the homogeneous boundary conditions (1.3), (1.4) :

$$(2.1)\quad \begin{cases} \dfrac{1}{2}\dfrac{d}{dt}\displaystyle\int_\Omega \rho^\varepsilon |\frac{\partial u^\varepsilon}{\partial t}|^2\,dx + \frac{1}{2}\frac{d}{dt}\int_\Omega A^\varepsilon\, grad\, u^\varepsilon\, grad\, u^\varepsilon\,dx \\ -\displaystyle\int_\Omega \gamma^\varepsilon\,\tau^\varepsilon\, grad\,\frac{\partial u^\varepsilon}{\partial t}\,dx = \int_\Omega f\,\frac{\partial u^\varepsilon}{\partial t}\,dx \end{cases}$$

$$(2.2)\quad \begin{cases} \dfrac{1}{2}\dfrac{d}{dt}\displaystyle\int_\Omega \beta^\varepsilon|\tau^\varepsilon|^2\,dx + \int_\Omega K^\varepsilon\, grad\,\tau^\varepsilon\, grad\,\tau^\varepsilon\,dx \\ +\displaystyle\int_\Omega \gamma^\varepsilon\, grad\,\frac{\partial u^\varepsilon}{\partial t}\,\tau^\varepsilon\,dx = < g\,,\tau^\varepsilon > \end{cases}$$

where $<,>$ denotes the duality pairing between $H^{-1}(\Omega)$ and $H_0^1(\Omega)$. Addition of these two equations miraculously eliminates the term $\int_\Omega \gamma^\varepsilon\,\tau^\varepsilon grad\,(\partial u^\varepsilon/\partial t)\,dx$ and integration in time yields

$$(2.3)\quad \begin{cases} \dfrac{1}{2}\displaystyle\int_\Omega [\rho^\varepsilon\,|\frac{\partial u^\varepsilon}{\partial t}|^2 + A^\varepsilon\, grad\, u^\varepsilon\, grad\, u^\varepsilon + \beta^\varepsilon\,|\tau^\varepsilon|^2]\,(x,t)dx \\ +\displaystyle\int_0^t\int_\Omega K^\varepsilon\, grad\,\tau^\varepsilon\, grad\,\tau^\varepsilon\,dx\,ds \\ =\dfrac{1}{2}\displaystyle\int_\Omega [\rho^\varepsilon\,|b^\varepsilon|^2 + A^\varepsilon\, grad\, a^\varepsilon\, grad\, a^\varepsilon + \beta^\varepsilon\,|c^\varepsilon|^2]\,dx \\ +\displaystyle\int_0^t\int_\Omega f\frac{\partial u^\varepsilon}{\partial t}\,dx\,ds + \int_0^t < g\,,\tau^\varepsilon >\,ds. \end{cases}$$

In view of the coercivity and boundedness assumptions (1.8)-(1.12), (2.3) implies that

$$(2.4)\quad \begin{cases} u^\varepsilon & \text{is bounded in} & L^\infty(0,T;H_0^1(\Omega)) \\ \dfrac{\partial u^\varepsilon}{\partial t} & \text{is bounded in} & L^\infty(0,T;L^2(\Omega)) \\ \tau^\varepsilon & \text{is bounded in} & L^\infty(0,T;L^2(\Omega))\cap L^2(0,T;H_0^1(\Omega)) \end{cases}$$

independently of ε.

This formal remark is at the root of the following existence and uniqueness result :

THEOREM 2.1. *Let the assumptions (1.8)-(1.12) be satisfied. Then for any fixed ε there exists a solution $(u^\varepsilon\,,\,\tau^\varepsilon)$*

$$(2.5)\quad \begin{cases} u^\varepsilon \in C^0(0,T;H_0^1(\Omega)) \\ \dfrac{\partial u^\varepsilon}{\partial t} \in C^0(0,T;L^2(\Omega)) \\ \tau^\varepsilon \in C^0(0,T;L^2(\Omega))\cap L^2(0,T;H_0^1(\Omega)) \end{cases}$$

of the system (1.1)-(1.7). Uniqueness holds in the (larger) class defined by (2.4). The solution satisfies equality (2.3) for any t in $[0,T]$. Moreover the estimate (2.4) holds true independently of ε. □

REMARK 2.1. Note that the regularity (2.5) (and even the "L^∞ in time" regularity defined by (2.4)) is sufficient for each term in (1.1), (1.2) to be meaningful. For example $\gamma^\varepsilon \, grad(\partial u^\varepsilon/\partial t)$ can be written as $\partial/\partial t\,(\gamma^\varepsilon \, grad\, u^\varepsilon)$ which has a distributional meaning since γ^ε belongs to $(L^\infty(\Omega))^N$ and $grad\, u^\varepsilon$ to $L^\infty(0,T;(L^2(\Omega))^N)$; the same is true for the other terms.

On the other hand, the sense of the term $\int_\Omega \gamma^\varepsilon \tau^\varepsilon \, grad(\partial u^\varepsilon/\partial t)\, dx$ is not clear in the above formal computation (2.1), (2.2). Nevertheless equality (2.3) (where every term has a meaning) holds true for the solution $(u^\varepsilon\,,\,\tau^\varepsilon)$ of the system. □

The proof of Theorem 2.1 can be considered as routine work, even if it is at times rather delicate : see [4], [6], [9] for a proof using semigroup theory (in which case one has to carefully examine the meaning of the equations, the density of the domain of the operators, and the question of uniqueness) and [2] for a proof by Galerkin's method (in which case one has to cautiously track the sense of the initial conditions, the continuous character of u^ε , $\partial u^\varepsilon/\partial t$ and τ^ε, and the question of uniqueness).

When the regularity (2.5) is met by $(u^\varepsilon\,,\,\partial u^\varepsilon/\partial t\,,\,\tau^\varepsilon)$ the sense of the initial conditions (1.5), (1.6), (1.7) is clear. We shall see in the following Proposition that these initial conditions are equally meaningful under the weaker hypothesis (2.4) whenever $(u^\varepsilon\,,\,\tau^\varepsilon)$ is a solution of (1.1), (1.2). In this statement $C^0_s(0,T;X)$ is the space of functions v such that the real function $t \to < h, v(t) >_{X',X}$ is continuous on $[0,T]$ for any $h \in X'$. This Proposition will be used in Section 4.

PROPOSITION 2.1. *Consider $(u^\varepsilon\,,\,\tau^\varepsilon)$ which belongs to the class "L^∞ in time" defined by (2.4) and is a solution of (1.1), (1.2) in a distributional sense. Then for any fixed ε*

$$(2.6)\qquad \begin{cases} u^\varepsilon \in C^0_s(0,T;H^1_0(\Omega)) \\ \dfrac{\partial u^\varepsilon}{\partial t} \in C^0_s(0,T;L^2(\Omega)) \\ \tau^\varepsilon \in C^0_s(0,T;L^2(\Omega)) \end{cases}$$

which lends a meaning to the initial conditions (1.5), (1.6), (1.7). Moreover if the entropy s^ε is defined as

$$(2.7)\qquad s^\varepsilon = \beta^\varepsilon \tau^\varepsilon + \gamma^\varepsilon \, grad\, u^\varepsilon$$

then

$$(2.8)\qquad \begin{cases} u^\varepsilon \in C^0(0,T;L^2(\Omega)) \\ \rho^\varepsilon \dfrac{\partial u^\varepsilon}{\partial t} \in C^0(0,T;H^{-1}(\Omega)) \\ s^\varepsilon \in C^0(0,T;H^{-1}(\Omega)). \end{cases}$$

Finally the initial conditions

$$u^\varepsilon(x,0) = a^\varepsilon(x) \quad in \quad \Omega \tag{2.9}$$

$$(\rho^\varepsilon \frac{\partial u^\varepsilon}{\partial t})(x,0) = \rho^\varepsilon(x)\, b^\varepsilon(x) \quad in \quad \Omega \tag{2.10}$$

$$s^\varepsilon(x,0) = \beta^\varepsilon(x)\, c^\varepsilon(x) + \gamma^\varepsilon(x)\, grad\, a^\varepsilon(x) \quad in \quad \Omega \tag{2.11}$$

are equivalent to the initial conditions (1.5), (1.6), (1.7). □

PROOF OF PROPOSITION 2.1. Recall that if X and Y are two Banach spaces such that X is reflexive and continuously imbedded in Y, then

$$L^\infty(0,T;X) \cap C^0(0,T;Y) \subset C^0_s(0,T;X)$$

(see e.g. [8] lemme 8.1 p. 297). In view of (2.4), u^ε belongs to $L^\infty(0,T;H^1_0(\Omega))$ with $\partial u^\varepsilon/\partial t$ in $L^\infty(0,T;L^2(\Omega))$, and thus it belongs to $C^0_s(0,T;H^1_0(\Omega))$.

By virtue of (2.4) together with equation (1.1), $\rho^\varepsilon\,(\partial u^\varepsilon/\partial t)$ belongs to $L^\infty(0,T;L^2(\Omega))$ with $\partial/\partial t\,(\rho^\varepsilon\,(\partial u^\varepsilon/\partial t))$ in $L^2(0,T;H^{-1}(\Omega))$. Thus $\rho^\varepsilon\,(\partial u^\varepsilon/\partial t)$ belongs to $C^0(0,T;H^{-1}(\Omega))$ and to $C^0_s(0,T;L^2(\Omega))$, which implies that

$$t \rightarrow \int_\Omega h \frac{\partial u^\varepsilon}{\partial t}\, dx = \int_\Omega (\frac{h}{\rho^\varepsilon})(\rho^\varepsilon \frac{\partial u^\varepsilon}{\partial t})\, dx \quad \text{is continuous on } [0,T]$$

for any $h \in L^2(\Omega)$. Thus $\partial u^\varepsilon/\partial t$ belongs to $C^0_s(0,T;L^2(\Omega))$, the initial condition on $\rho^\varepsilon\,(\partial u^\varepsilon/\partial t)$ is meaningful, (2.10) is satisfied and is exactly (and not only formally) equivalent to (1.6) whenever $(u^\varepsilon\,,\,\tau^\varepsilon)$ belongs to the class "L^∞ in time" defined by (2.4). Note that this equivalence is straightforward when the continuity (2.5) holds true.

Finally let us consider the entropy s^ε defined by (2.7). From (2.4) and equation (1.2) we deduce that s^ε belongs to $L^\infty(0,T;L^2(\Omega))$ with $\partial s^\varepsilon/\partial t$ in $L^2(0,T;H^{-1}(\Omega))$. Thus s^ε belongs to $C^0(0,T;H^{-1}(\Omega))$, which gives a sense to the initial condition on s^ε. On the other hand s^ε belongs to $C^0_s(0,T;L^2(\Omega))$. Since u^ε lies in $C^0_s(0,T;H^1_0(\Omega))$ the function

$$\begin{cases} t \rightarrow \int_\Omega h\,\beta^\varepsilon\,\tau^\varepsilon\, dx = \int_\Omega h\, s^\varepsilon\, dx - \int_\Omega h\,\gamma^\varepsilon\, grad\, u^\varepsilon\, dx \\ \text{is continuous on } [0,T] \text{ for any } h \in L^2(\Omega). \end{cases} \tag{2.12}$$

Taking $h' = h/\beta^\varepsilon$ for test function in (2.12) proves that τ^ε belongs to $C^0_s(0,T;L^2(\Omega))$. Finally (2.12) also proves that (2.11) holds true and that (1.5), (1.7) is equivalent to (2.9), (2.11) whenever $(u^\varepsilon\,,\,\tau^\varepsilon)$ belongs to the class "L^∞ in time" defined by (2.4). Note that this equivalence is straightforward when the continuity (2.5) holds true. □

3. STATEMENT OF THE HOMOGENIZATION RESULT.

Consider a sequence of data which meet the assumptions (1.8)-(1.12). In this setting, Theorem 2.1 asserts that the unique solution $(u^\varepsilon\,,\,\tau^\varepsilon)$ of the thermoelasticity problem (1.1)-(1.7) is bounded as ε tend to 0 (see (2.4)). Our goal in this Section is to pass to the limit in ε.

THEOREM 3.1. *If the assumptions (1.8)-(1.12) are satisfied, there exists a subsequence $(u^\varepsilon, \tau^\varepsilon)$ (still indexed by the same superscript ε), which converges to (u, τ) in the following sense*

$$(3.0)\qquad \begin{cases} u^\varepsilon \rightharpoonup u \text{ weak } * \text{ in } L^\infty(0,T;H_0^1(\Omega)) \\ \dfrac{\partial u^\varepsilon}{\partial t} \rightharpoonup \dfrac{\partial u}{\partial t} \text{ weak } * \text{ in } L^\infty(0,T;L^2(\Omega)) \\ \tau^\varepsilon \rightharpoonup \tau \text{ weak } * \text{ in } L^\infty(0,T;L^2(\Omega)) \cap L^2(0,T;H_0^1(\Omega)). \end{cases}$$

In (3.0) (u, τ) is the unique solution (given by Theorem 2.1) of the following homogenized thermoelasticity system :

$$(3.1)\qquad \bar{\rho}\,\frac{\partial^2 u}{\partial t^2} - div(A^0\, grad\, u - \gamma^0\,\tau) = f \quad in \quad \Omega \times (0,T)$$

$$(3.2)\qquad (\bar{\beta} + \kappa)\,\frac{\partial \tau}{\partial t} - div(K^0\, grad\, \tau) + \gamma^0\, grad\, \frac{\partial u}{\partial t} = g \quad in \quad \Omega \times (0,T)$$

$$(3.3)\qquad u(x,t) = 0 \quad on \quad \partial\Omega \times (0,T)$$

$$(3.4)\qquad \tau(x,t) = 0 \quad on \quad \partial\Omega \times (0,T)$$

$$(3.5)\qquad u(x,0) = a^0(x) \quad in \quad \Omega$$

$$(3.6)\qquad \frac{\partial u}{\partial t}(x,0) = b^0(x) \quad in \quad \Omega$$

$$(3.7)\qquad \tau(x,0) = c^0(x) \quad in \quad \Omega$$

where the definitions of $\bar{\rho}$, $\bar{\beta}$, A^0, K^0, κ, a^0, b^0 and c^0 are given below. □

In view of the bound (1.8), we are at liberty to extract a subsequence ε' such that

$$(3.8)\qquad \begin{cases} \rho^{\varepsilon'} \rightharpoonup \bar{\rho}\,,\ \beta^{\varepsilon'} \rightharpoonup \bar{\beta} \text{ weak } * \text{ in } L^\infty(\Omega) \\ \text{with } \ \lambda_1 \le \bar{\rho} \le \lambda_2\,,\ \lambda_1 \le \bar{\beta} \le \lambda_2 \text{ a.e. in } \Omega\,; \end{cases}$$

this defines $\bar{\rho}$ and $\bar{\beta}$.

The matrix A^0 (and in a similar manner the matrix K^0) is defined as the homogenized matrix obtained from a subsequence of A^ε. It is now well known (see [11], [16], [17] and also [1], [14] in a periodic setting) that one can extract a subsequence ε' such that for a certain matrix A^0 which satisfies the same inequality

(1.9) as A^ε, the following property holds true : for any sequence h^ε converging strongly to h in $H^{-1}(\Omega)$, the solutions $v^{\varepsilon'}$ of the elliptic problem

$$(3.9a) \qquad -div(A^{\varepsilon'}\, grad\, v^{\varepsilon'}) = h^{\varepsilon'} \text{ in } \Omega\,,\ v^\varepsilon \in H_0^1(\Omega).$$

satisfy

$$(3.9b) \qquad A^{\varepsilon'}\, grad\, v^{\varepsilon'} \rightharpoonup A^0\, grad\, v \text{ weakly in } (L^2(\Omega))^N$$

for v the unique solution of

$$(3.9c) \qquad -div(A^0\, grad\, v) = h \text{ in } \Omega\,,\ v \in H_0^1(\Omega).$$

If (3.9) holds true for any strongly converging sequence $h^{\varepsilon'}$ in $H^{-1}(\Omega)$, the sequence $A^{\varepsilon'}$ is said to $H-converge$ to A^0. The matrix A^0 represents in essence the effective medium associated to the heterogeneous medium A^ε.

The matrices A^0 and K^0 are thus defined.

For a sequence $A^{\varepsilon'}$ which H-converges to A^0, there exist N functions $w_i^{\varepsilon'}$ in $H^1(\Omega)$ $(1 \le i \le N)$ such that

$$(3.10a) \qquad \begin{cases} -\, div(A^{\varepsilon'}\, grad\, w_i^{\varepsilon'}) = -div(A^0\, e_i) \text{ in } \Omega \\ w_i^{\varepsilon'} \rightharpoonup (e_i\,,\, x) = x_i \text{ weakly in } H^1(\Omega) \end{cases}$$

where e_i denotes the i-th vector of some orthonormal basis of $\mathbb{R}^N$ (to define such w_i^ε it is for example sufficient to impose a convenient boundary condition on $\partial\Omega$, as in the proof of (3.13) below). We define the matrix $P^{\varepsilon'} \in (L^2(\Omega))^{N^2}$ by

$$(3.10b) \qquad P^{\varepsilon'}\, e_i = grad\, w_i^{\varepsilon'} \text{ in } \Omega.$$

If $v^{\varepsilon'}$ is the solution of (3.9) then ([11], [17])

$$(3.10c) \qquad \begin{cases} grad\, v^{\varepsilon'} = P^{\varepsilon'}\, grad\, v + r^{\varepsilon'} \\ \text{with } r^{\varepsilon'} \to 0 \text{ strongly in } (L^1(\Omega))^N. \end{cases}$$

This result further specifies the structure of the solution of (3.9a). The matrix $P^{\varepsilon'}$ is called the *corrector matrix associated to* $A^{\varepsilon'}$.

From the matrix $P^{\varepsilon'}$ we can now define γ^0 by extracting a subsequence (still indexed by ε') such that

$$(3.11) \qquad {}^tP^{\varepsilon'}\, \gamma^{\varepsilon'} \rightharpoonup \gamma^0 \text{ weakly in } (L^2(\Omega))^N.$$

Note that ${}^tP^{\varepsilon'}\, \gamma^{\varepsilon'}$ is bounded in $(L^2(\Omega))^N$ in view of (3.10a) and (1.10). It can be proved ([11],[17] ; in a similar spirit, see the proof of (3.15) below) that γ^0 inherits from the L^∞ bound on γ^ε the regularity property

$$(3.12) \qquad \gamma^0 \in (L^\infty(\Omega))^N.$$

Let $y^{\varepsilon'}$ be any function such that

$$(3.13) \qquad \begin{cases} -\,div(A^{\varepsilon'}\,grad\,y^{\varepsilon'} - (\gamma^{\varepsilon'} - \gamma^0)) = 0 \text{ in } \Omega \\ y^{\varepsilon'} \rightharpoonup 0 \text{ weakly in } H_0^1(\Omega) \end{cases}$$

(we will prove at the end of this Section that such a sequence exists). We define κ by extracting a new subsequence (still indexed by ε') such that

$$(3.14) \qquad \gamma^{\varepsilon'}\,grad\,y^{\varepsilon'} \rightharpoonup \kappa \text{ weakly in } L^2(\Omega).$$

It will be proved at the end of this Section that

$$(3.15) \qquad \kappa \in L^\infty(\Omega) \quad \text{and} \quad \kappa \geq 0 \text{ a.e. in } \Omega.$$

The initial conditions a^0, b^0, and c^0 have yet to be defined. In view of (1.12) we can extract a subsequence (still indexed by ε') such that

$$(3.16) \qquad a^{\varepsilon'} \rightharpoonup a^0 \text{ weakly in } H_0^1(\Omega)$$

which defines a^0 ; we can also extract a subsequence such that

$$(3.17) \qquad \rho^{\varepsilon'}\,b^{\varepsilon'} \rightharpoonup \overline{\rho b} \text{ weakly in } L^2(\Omega).$$

From $\overline{\rho b}$ and $\overline{\rho}$ defined in (3.8) we define b^0 by

$$(3.18) \qquad b^0 = \overline{\rho b}\,/\,\overline{\rho} \in L^2(\Omega)$$

(note that $1/\overline{\rho} \in L^\infty(\Omega)$).

The initial value c^0 of the temperature field τ is obtained with the help of the initial value $s^\varepsilon(x,0)$ of the entropy (see (2.11)). Defining d^ε as

$$(3.19a) \qquad d^\varepsilon = \beta^\varepsilon\,c^\varepsilon + \gamma^\varepsilon\,grad\,a^\varepsilon$$

we extract a subsequence (still indexed by ε') such that

$$(3.19b) \qquad d^{\varepsilon'} \rightharpoonup d^0 \text{ weakly in } L^2(\Omega).$$

Then c^0 is defined by the formula

$$(3.20) \qquad (\overline{\beta} + \kappa)\,c^0 + \gamma^0\,grad\,a^0 = d^0$$

which in view of the previous results (and particularly of (3.15) which yields $(\overline{\beta} + \kappa)^{-1} \in L^\infty(\Omega)$) implies that

$$(3.21) \qquad c^0 \in L^2(\Omega).$$

All quantities which appear in the homogenized system (3.1)-(3.7) are now defined.

REMARK 3.1. In a periodic setting where ρ^ε, β^ε, A^ε, K^ε and γ^ε are given by formulas like $\rho^\varepsilon(x) = \rho(x/\varepsilon)$ with ρ periodic, the quantities $\bar\rho$, $\bar\beta$, A^0, K^0, γ^0 and κ are constants (scalars, matrices or vectors) and are given by "explicit" formulas involving the solutions of partial differential equations with periodic solutions : see [4],[6]. □

REMARK 3.2. In the special case where the initial conditions of the system to be homogenized are fixed, namely

$$a^\varepsilon = a\ ,\ b^\varepsilon = b\ ,\ c^\varepsilon = c \text{ with } a \in H_0^1(\Omega)\ ,\ b \in L^2(\Omega)\ ,\ c \in L^2(\Omega)$$

it is easily seen that

$$a^0 = a \quad , \quad b^0 = b.$$

Thus there is no change in the initial conditions for u and $\partial u/\partial t$.

A completely different phenomenon occurs for the initial condition on τ : indeed, if for some subsequence still indexed by ε' one has

$$\gamma^{\varepsilon'} \rightharpoonup \bar\gamma \text{ weakly in} \,(L^2(\Omega))^N,$$

it is easily seen from (3.19), (3.20) that the initial condition c^0 on τ is given by

$$c^0 = \frac{1}{\bar\beta + \kappa}\,[\bar\beta\, c + (\bar\gamma - \gamma^0)\, grad\, a] \tag{3.22}$$

which is in general different from c. □

REMARK 3.3. The homogenized system (3.1)-(3.7) ressembles the initial system (1.1)-(1.7). Two differences have to be pointed out :

(i) the presence of κ in the coefficient of $\partial\tau/\partial t$; we will see (Remark 4.1 below) that the presence of such a coefficient is a consequence of the hidden presence of τ in the term $\gamma^\varepsilon\, grad\, u^\varepsilon$.

(ii) the change in the initial condition for τ (see Remark 3.2 above) ; we will see in Section 4 that, as far as the initial conditions are concerned, the entropy s^ε (and not the temperature τ^ε) is the quantity to investigate.

The most surprising feature of the homogenization result does not reside however in the above described phenomena but rather in the very fact that the homogenized system is so close to the original one. Such is not generally the case when dealing with systems with time-dependent effects, see e.g. [7], [10], [14], [19]. □

PROOF OF (3.13). We prove that there exists a sequence $y^{\varepsilon'}$ which satisfies (3.13). Assume that $y^{\varepsilon'}$ satisfies

$$-div(A^{\varepsilon'}\, grad\, y^{\varepsilon'} - (\gamma^{\varepsilon'} - \gamma^0)) = 0 \tag{3.23}$$

and a convenient boundary condition implying that $y^{\varepsilon'}$ is bounded in $H^1(\Omega)$. Define

$$\zeta^{\varepsilon'} = A^{\varepsilon'}\, grad\, y^{\varepsilon'} - (\gamma^{\varepsilon'} - \gamma^0). \tag{3.24}$$

Extracting a subsequence (indexed by ε") we may assume that

$$\left\{\begin{array}{l} grad\, y^{\varepsilon''} \rightharpoonup grad\, y \text{ weakly in } (L^2(\Omega))^N \\ \zeta^{\varepsilon''} \rightharpoonup \zeta \text{ weakly in } (L^2(\Omega))^N. \end{array}\right. \tag{3.25}$$

Since $\zeta^{\varepsilon''}$ is divergence free, application of the div-curl lemma (see [12], [18]) to $(\zeta^{\varepsilon''}, grad\, w_i^{\varepsilon''})$ (see (3.10)) yields

$$(A^{\varepsilon''} grad\, y^{\varepsilon''} - (\gamma^{\varepsilon''} - \gamma^0))\, grad\, w_i^{\varepsilon''} \rightharpoonup \zeta e_i \text{ in } \mathcal{D}'(\Omega)$$

which in view of the definition (3.11) of γ^0 implies that

$$A^{\varepsilon''} grad\, y^{\varepsilon''} grad\, w_i^{\varepsilon''} \rightharpoonup \zeta e_i \text{ in } \mathcal{D}'(\Omega).$$

A new application of the div-curl lemma to $(A^{\varepsilon''} grad\, w_i^{\varepsilon''}, grad\, y^{\varepsilon''})$ yields $A^0\, e_i\, grad\, y = \zeta\, e_i$, i.e. $\zeta = A^0\, grad\, y$. Since $-div\, \zeta = 0$ we obtain $-div\, A^0\, grad\, y = 0$, which implies that $y = 0$, thus $\zeta = 0$ provided that the boundary conditions on $y^{\varepsilon'}$ are chosen in a manner such that the trace of y is zero on $\partial\Omega$. Such is the case whenever $y^{\varepsilon'} = 0$ on $\partial\Omega$, but other choices could be more convenient, for example $y^{\varepsilon'}$ periodic when $A^\varepsilon(x) = A(x/\varepsilon)$ with A periodic.

Since, under these assumptions, $y = 0$ and $\zeta = 0$, the convergence (3.25) takes place for the whole sequence ε' ; the existence of $y^{\varepsilon'}$ satisfying (3.13) is established.

PROOF OF (3.15). A new application of the div-curl lemma to $(\zeta^{\varepsilon'}, grad\, y^{\varepsilon'})$ where $\zeta^{\varepsilon'}$ is defined by (3.24) yields

$$(A^{\varepsilon'} grad\, y^{\varepsilon'} - (\gamma^{\varepsilon'} - \gamma^0))\, grad\, y^{\varepsilon'} \rightharpoonup 0 \text{ in } \mathcal{D}'(\Omega)$$

which in view of the definition (3.14) of κ implies that

$$A^{\varepsilon'} grad\, y^{\varepsilon'} grad\, y^{\varepsilon'} \rightharpoonup \kappa \text{ in } \mathcal{D}'(\Omega) \tag{3.26}$$

since $grad\, y^{\varepsilon'} \rightharpoonup 0$.

This immediately implies that $\kappa \geq 0$ in the sense of measures, and thus almost everywhere on Ω because κ belongs to $L^2(\Omega)$.

Let us finally prove that κ belongs to $L^\infty(\Omega)$. For any $\varphi \in \mathcal{D}(\Omega)$ we have

$$\begin{aligned} \Big|\int_\Omega \gamma^{\varepsilon'} grad\, y^{\varepsilon'} \varphi^2\, dx\Big| &\leq \|\gamma^{\varepsilon'} \varphi\|_{(L^2(\Omega))^N} \|grad\, y^{\varepsilon'} \varphi\|_{(L^2(\Omega))^N} \\ &\leq \|\gamma^{\varepsilon'}\|_{(L^\infty(\Omega))^N} \|\varphi\|_{L^2(\Omega)} \frac{1}{\lambda_1}\Big(\int_\Omega A^{\varepsilon'} grad\, y^{\varepsilon'} grad\, y^{\varepsilon'} \varphi^2\, dx\Big)^{1/2}. \end{aligned}$$

Using the bound (1.10) on γ^ε and passing to the limit with the help of (3.14), (3.26) yields

$$\Big|\int_\Omega \kappa\varphi^2\, dx\Big| \leq \frac{\lambda_2}{\lambda_1} \|\varphi\|_{L^2(\Omega)} \Big(\int_\Omega \kappa\varphi^2\, dx\Big)^{1/2},$$

i.e.

$$\int_\Omega \kappa\, \varphi^2\, dx \le \left(\frac{\lambda_2}{\lambda_1}\right)^2 \int_\Omega \varphi^2\, dx \qquad \forall \varphi \in \mathcal{D}(\Omega).$$

Since κ is non negative this implies that κ belongs to $L^\infty(\Omega)$ with $\|\sigma\|_{L^\infty(\Omega)} \le (\lambda_2/\lambda_1)^2$. Thus (3.15) is proved. □

4. PROOF OF THE HOMOGENIZATION RESULT.

We have defined in Section 3 all the quantities which appear in the homogenized system (3.1)-(3.7). Let ε' denote a subsequence such that $A^{\varepsilon'}$ and $K^{\varepsilon'}$ H-converge (see (3.9)) to A^0 and K^0 and such that all the convergences (3.8), (3.11), (3.14), (3.16), (3.17) and (3.19) take place. The statement of Theorem 3.1 is now proved to hold true for this subsequence. The proof consists in extracting a new subsequence $(u^{\varepsilon''}, \tau^{\varepsilon''})$ which converges weakly to some (u, τ) in the topology given by (3.0) which will be shown to be a solution of (3.1)-(3.7). Thus (u, τ) is uniquely determined, and the whole sequence $(u^{\varepsilon'}, \tau^{\varepsilon'})$ converges to (u, τ).

For the sake of notational simplicity the subsequence ε' is from now onward denoted by ε.

4.1. Passing to the limit in the equations.

In order to pass to the limit in the equations (1.1) and (1.2) we transform these evolution equations into elliptic equations, which could be performed with the help of the Laplace transform (see [4], [6]). Our preference goes here to the introduction of the following auxiliary functions defined for a given $\varphi \in \mathcal{D}(0,T)$ by :

$$(4.1)\qquad \begin{cases} \hat{u}^\varepsilon(x) = \displaystyle\int_0^T u^\varepsilon(x,t)\,\varphi(t)\,dt \\ \hat{\tau}^\varepsilon(x) = \displaystyle\int_0^T \tau^\varepsilon(x,t)\,\varphi(t)\,dt. \end{cases}$$

In view of (3.0) we obtain

$$(4.2)\qquad \begin{cases} \hat{u}^\varepsilon \rightharpoonup \hat{u} = \displaystyle\int_0^T u(\cdot,t)\,\varphi(t)\,dt \text{ weakly in } H_0^1(\Omega) \text{ and strongly in } L^2(\Omega) \\ \hat{\tau}^\varepsilon \rightharpoonup \hat{\tau} = \displaystyle\int_0^T \tau(\cdot,t)\,\varphi(t)\,dt \text{ weakly in } H_0^1(\Omega) \text{ and strongly in } L^2(\Omega). \end{cases}$$

On the other hand it is deduced from (1.1) that

$$(4.3)\qquad -div(A^\varepsilon\, grad\, \hat{u}^\varepsilon - \gamma^\varepsilon\, \hat{\tau}^\varepsilon) = \int_0^T f\,\varphi\, dt - \rho^\varepsilon \int_0^T u^\varepsilon \frac{\partial^2 \varphi}{\partial t^2}\, dt.$$

Define $\xi^\varepsilon = A^\varepsilon\, grad \hat{u}^\varepsilon - \gamma^\varepsilon\, \hat{\tau}^\varepsilon$ and extract a subsequence such that

$$\xi^\varepsilon \rightharpoonup \xi \text{ weakly in } (L^2(\Omega))^N.$$

Since the right hand side of (4.3) is compact in $H^{-1}(\Omega)$, application of the div-curl lemma ([12], [18]) to $(\xi^\varepsilon, grad\, w_i^\varepsilon)$ (see (3.10) for the definition of w_i^ε) yields

$$(A^\varepsilon\, grad\, \hat{u}^\varepsilon - \gamma^\varepsilon\, \hat{\tau}^\varepsilon) grad\, w_i^\varepsilon \rightharpoonup \xi\, e_i \text{ in } \mathcal{D}'(\Omega). \tag{4.4a}$$

Another application of the div-curl lemma to $(A\, grad\, w_i^\varepsilon, grad\, \hat{u}^\varepsilon)$ yields

$$A^\varepsilon\, grad\, \hat{u}^\varepsilon\, grad\, w_i^\varepsilon = A^\varepsilon\, grad\, w_i^\varepsilon\, grad\, \hat{u}^\varepsilon \rightharpoonup A^0\, e_i\, grad\, \hat{u} \text{ in } \mathcal{D}'(\Omega) \tag{4.4b}$$

while the definition (3.11) of γ^0 and the strong $L^2(\Omega)$ convergence of $\hat{\tau}^\varepsilon$ imply that

$$\gamma^\varepsilon\, \hat{\tau}^\varepsilon\, grad\, w_i^\varepsilon = {}^tP^\varepsilon\, \gamma^\varepsilon\, e_i\, \hat{\tau}^\varepsilon \rightharpoonup \gamma^0\, e_i\, \hat{\tau} \text{ in } \mathcal{D}'(\Omega). \tag{4.4c}$$

Thus $\xi = A^0\, grad\, \hat{u} - \gamma^0\, \hat{\tau}$. Passing to the limit in (4.3) allows to recover (3.1) in the distributional sense since $\varphi \in \mathcal{D}(0,T)$ is arbitrary.

Passing to the limit in (1.2) is more delicate. Through the above described procedure (1.2) becomes

$$\begin{aligned} -div(K^\varepsilon\, grad\, \hat{\tau}^\varepsilon) = &\int_0^T g\, \varphi\, dt + \beta^\varepsilon \int_0^T \tau^\varepsilon \frac{\partial \varphi}{\partial t}\, dt \\ &+ \gamma^\varepsilon\, grad\Big(\int_0^T u^\varepsilon \frac{\partial \varphi}{\partial t}\, dt\Big). \end{aligned} \tag{4.5}$$

Let us first study the last term of the right hand side of (4.5). Set

$$\breve{u}^\varepsilon = \int_0^T u^\varepsilon \frac{\partial \varphi}{\partial t}\, dt \quad , \quad \breve{\tau}^\varepsilon = \int_0^T \tau^\varepsilon \frac{\partial \varphi}{\partial t}\, dt.$$

Then $(\breve{u}^\varepsilon, \breve{\tau}^\varepsilon)$ converges weakly in $(H_0^1(\Omega))^2$ to $(\breve{u}, \breve{\tau})$ and satisfies

$$-div(A^\varepsilon\, grad\, \breve{u}^\varepsilon - \gamma^\varepsilon\, \breve{\tau}^\varepsilon) = \int_0^T f \frac{\partial \varphi}{\partial t}\, dt - \rho^\varepsilon \int_0^T u^\varepsilon \frac{\partial^3 \varphi}{\partial t^3}\, dt. \tag{4.6}$$

Further

$$\begin{aligned} \gamma^\varepsilon\, grad\, \breve{u}^\varepsilon = &(\gamma^\varepsilon - \gamma^0 - A^\varepsilon\, grad\, y^\varepsilon)\, grad\, \breve{u}^\varepsilon \\ &+ (A^\varepsilon\, grad\, \breve{u}^\varepsilon - \gamma^\varepsilon\, \breve{\tau}^\varepsilon)\, grad\, y^\varepsilon \\ &+ \gamma^0\, grad\, \breve{u}^\varepsilon + \gamma^\varepsilon\, grad\, y^\varepsilon\, \breve{\tau}^\varepsilon. \end{aligned} \tag{4.7}$$

Application of the div-curl lemma implies that the first two terms converge to 0 in $\mathcal{D}'(\Omega)$; definition (3.14) of κ together with the boundedness of $\gamma^\varepsilon\, grad\, \breve{u}^\varepsilon$ in $L^2(\Omega)$ yield

$$\gamma^\varepsilon\, grad\, \breve{u}^\varepsilon \rightharpoonup \gamma^0\, grad\, \breve{u} + \kappa\, \breve{\tau} \text{ weakly in } L^2(\Omega). \tag{4.8}$$

It is now easy to pass to the limit in (4.5) using the definition (3.9) of H-convergence. Since φ is arbitrary in $\mathcal{D}(0,T)$, (3.2) is recovered.

REMARK 4.1. The convergence (4.8) of $\gamma^\varepsilon\, grad\, \check{u}^\varepsilon$ to $\gamma^0\, grad\, \check{u} + \kappa\, \check{\tau}$ is the source of the presence of κ in the coefficient of $\partial\tau/\partial t$ in (3.2). This phenomenon is better explained by application to equation (4.6) of the following corrector result, which proves that the displacement $grad\, \check{u}^\varepsilon$ is the sum of two parts, carrying $grad\, \check{u}$ and $\check{\tau}$ respectively. □

PROPOSITION 4.1. *Let v^ε be the solution of*

$$(4.9) \qquad \begin{cases} -\, div(A^\varepsilon\, grad\, v^\varepsilon - \gamma^\varepsilon\, \theta^\varepsilon) = h^\varepsilon \text{ in } \Omega \\ v^\varepsilon \in H_0^1(\Omega) \end{cases}$$

where the sequence A^ε satisfies (1.9) and H-converges (see (3.9)) to A^0, where γ^ε, γ^0 and κ satisfy (1.10), (3.11), (3.12) and (3.15), and where

$$\begin{aligned} \theta^\varepsilon &\to \theta \text{ strongly in } L^2(\Omega) \\ h^\varepsilon &\to h \text{ strongly in } H^{-1}(\Omega). \end{aligned}$$

Then

$$(4.10) \qquad \begin{cases} grad\, v^\varepsilon = P^\varepsilon\, grad\, v + \theta\, grad\, y^\varepsilon + r^\varepsilon \\ \text{with } r^\varepsilon \to 0 \text{ strongly in } (L^1(\Omega))^N \end{cases}$$

where v is the solution of

$$(4.11) \qquad -div(A^0\, grad\, v - \gamma^0\, \theta) = h \text{ in } \Omega,\ v \in H_0^1(\Omega).$$

□

REMARK 4.2. When additional regularity holds for v and θ, the strong convergence of r^ε in (4.10) takes place in a better space : for example when v belongs to $C^1(\bar{\Omega})$ and θ to $C^0(\bar{\Omega})$, one gets

$$r^\varepsilon \to 0 \text{ strongly in } (L^2(\Omega))^N.$$

The same result holds true if P^ε and $grad\, y^\varepsilon$ are bounded in $(L^\infty(\Omega))^{N^2}$ and $(L^\infty(\Omega))^N$ while v and θ merely belongs to $H_0^1(\Omega)$ and $L^2(\Omega)$ respectively. □

PROOF OF PROPOSITION 4.1. Application of the div-curl lemma as in (4.4) proves that the sequence v^ε tends to v in $H_0^1(\Omega)$ where v is the solution of (4.11). The analogue of decomposition (4.7) then proves that

$$(4.12) \qquad \gamma^\varepsilon\, grad\, v^\varepsilon \rightharpoonup \gamma^0\, grad\, v + \kappa\, \theta \text{ weakly in } (L^2(\Omega))^N.$$

Consider now for any given ϕ in $\mathcal{D}(\Omega)$ and ψ in $\mathcal{D}(\Omega)$
(4.13)

$$\begin{aligned} X^\varepsilon &= \int_\Omega A^\varepsilon [grad\, v^\varepsilon - P^\varepsilon\, grad\, \phi - \psi\, grad\, y^\varepsilon][grad\, v^\varepsilon - P^\varepsilon\, grad\, \phi - \psi\, grad\, y^\varepsilon]\, dx \\ &= I^\varepsilon + II^\varepsilon \end{aligned}$$

where

$$I^\varepsilon = \int_\Omega [A^\varepsilon\, grad\, v^\varepsilon - \gamma^\varepsilon\, \theta^\varepsilon - A^\varepsilon\, P^\varepsilon\, grad\, \phi - \psi\{A^\varepsilon\, grad\, y^\varepsilon - (\gamma^\varepsilon - \gamma_0)\}]$$
$$[grad\, v^\varepsilon - P^\varepsilon\, grad\, \phi - \psi\, grad\, y^\varepsilon]\, dx$$
$$II^\varepsilon = \int_\Omega [\gamma^\varepsilon(\theta^\varepsilon - \psi) + \gamma^0\, \psi]\, [grad\, v^\varepsilon - P^\varepsilon\, grad\, \phi - \psi\, grad\, y^\varepsilon]\, dx.$$

Since ϕ and ψ are smooth, application of the div-curl lemma in I^ε is licit, while (4.12), (3.11) and (3.14) give the value of the limit of II^ε ; we obtain

$$\begin{aligned} X^\varepsilon \to &\int_\Omega [A^0\, grad\, v - \gamma^0\, \theta - A^0\, grad\, \phi - \psi\, 0][grad\, v - grad\, \phi - \psi\, 0]\, dx \\ &+ \int_\Omega (\theta - \psi)\, [\gamma^0\, grad\, v + \kappa\, \theta - \gamma^0\, grad\, \phi - \kappa\, \psi]\, dx \\ &+ \int_\Omega \psi\, \gamma^0 [grad\, v - grad\, \phi - \psi\, 0]\, dx \\ &= \int_\Omega A^0\, grad(v - \phi)\, grad(v - \phi)\, dx + \int_\Omega \kappa(\theta - \psi)^2\, dx \end{aligned}$$

The definition (4.13) of X^ε implies that

$$\begin{aligned} &\limsup_{\varepsilon\to 0} \|grad\, v^\varepsilon - P^\varepsilon\, grad\, \phi - \psi\, grad\, y^\varepsilon\|^2_{(L^2(\Omega))^N} \\ &\le \frac{\lambda_2}{\lambda_1} \|grad(v - \phi)\|^2_{(L^2(\Omega))^N} + \frac{\|\kappa\|_{L^\infty(\Omega)}}{\lambda_1} \|\theta - \psi\|^2_{L^2(\Omega)}. \end{aligned} \tag{4.14}$$

The result (4.10) is then deduced from (4.14) upon setting

$$\begin{aligned} r^\varepsilon &= grad\, v^\varepsilon - P^\varepsilon\, grad\, v - \theta\, grad\, y^\varepsilon \\ &= grad\, v^\varepsilon - P^\varepsilon\, grad\, \phi - \psi\, grad\, y^\varepsilon - P^\varepsilon\, grad(v - \phi) - (\theta - \psi)\, grad\, y^\varepsilon \end{aligned} \tag{4.15}$$

and recalling that P^ε and $grad\, y^\varepsilon$ are bounded in $(L^2(\Omega))^{N^2}$ and $(L^2(\Omega))^N$. Relations (4.14) and (4.15) also imply the results stated in Remark 4.2. □

4.2. Passing to the limit in the initial conditions.

Recall that for X and Y Banach spaces with X compactly imbedded in Y, the set $V = \{v$ bounded in $L^\infty(0,T;X)$ with $\partial v/\partial t$ bounded in $L^2(0,T;Y)\}$ is relatively compact in $C^0(0,T;Y)$ (see e.g. [15] Corollary 4 p. 85).

In view of (3.0), u^ε, which is bounded in $L^\infty(0,T;H^1_0(\Omega))$ with $\partial u^\varepsilon/\partial t$ bounded in $L^\infty(0,T;L^2(\Omega))$, converges strongly to u in $C^0(0,T;L^2(\Omega))$. It is thus easy to pass to the limit in (1.5) and to obtain

$$u(0) = a^0 \text{ in } \Omega. \tag{4.16}$$

Similarly, $\rho^\varepsilon\, \partial u^\varepsilon/\partial t$ is bounded in $L^\infty(0,T;L^2(\Omega))$ while $\partial/\partial t\, (\rho^\varepsilon\, (\partial u^\varepsilon/\partial t))$ is bounded in $L^\infty(0,T;H^{-1}(\Omega))$, because of (2.4) and equation (1.1). This implies that $\rho^\varepsilon\, \partial u^\varepsilon/\partial t$ converges strongly in $C^0(0,T;H^{-1}(\Omega))$ to $\bar\rho\, \partial u/\partial t$. It is now possible to pass to the limit in (2.10). We obtain in view of (3.17)

$$\bar\rho\, \frac{\partial u}{\partial t}\,(x,0) = \overline{\rho\, b}. \tag{4.17}$$

But we proved in Proposition 2.1 that any solution $(u\,,\,\tau)$ of (3.1), (3.2) which belongs to the class "L^∞ in time" defined by (2.4) actually belongs to the class "C^0_s in time" defined by (2.6), and that (4.17) is equivalent to the initial condition $(\partial u/\partial t)\,(x,0) = \overline{\rho\, b}/\bar\rho$ which coincides with (3.6) in view of the definition (3.18) of b^0.

Finally consider the entropy defined by (2.7), namely

$$s^\varepsilon = \beta^\varepsilon\, \tau^\varepsilon + \gamma^\varepsilon\, grad\, u^\varepsilon \tag{4.18}$$

and the entropy s corresponding to the solution $(u\,,\,\tau)$ of (3.1)-(3.7), defined by

$$s = (\bar\beta + \kappa)\tau + \gamma^0\, grad\, u. \tag{4.19}$$

Note that s^ε is bounded in $L^\infty(0,T;L^2(\Omega))$. Multiplying (4.18) by φ in $\mathcal{D}(0,T)$ integrating on (0,T) and using an analogue of (4.8) allows to deduce that

$$s^\varepsilon \;\rightharpoonup\; s \text{ weak } * \text{ in } L^\infty(0,T;L^2(\Omega)). \tag{4.20}$$

On the other hand s^ε is bounded in $L^\infty(0,T;L^2(\Omega))$ with $\partial s^\varepsilon/\partial t$ bounded in $L^2(0,T;H^{-1}(\Omega))$ by virtue of (2.4) and equation (1.2). Thus

$$s^\varepsilon \;\rightarrow\; s \text{ strongly in } C^0(0,T;H^{-1}(\Omega)). \tag{4.21}$$

In view of the definitions (3.19), (3.20), this permits to pass to the limit in

$$s^\varepsilon(0) = d^\varepsilon \text{ in } \Omega \tag{4.22}$$

and to obtain

$$s(0) = d^0 = (\bar\beta + \kappa)\, c^0 + \gamma^0\, grad\, a^0 \text{ in } \Omega. \tag{4.23}$$

Since $(u\,,\,\tau)$ is a solution of (3.1), (3.2) which belongs to the class "L^∞ in time" defined in (2.4), Proposition 2.1 proves that the initial conditions (4.16), (4.23) are equivalent to (3.5), (3.7).

The uniqueness result of Theorem 2.1 completes the proof of Theorem 3.1. □

5. STATEMENT OF THE CORRECTOR RESULT.

This Section is devoted to the study of the structure of the solution $(u^\varepsilon\,,\,\tau^\varepsilon)$ when ε tends to zero. It follows closely the method used in the study of the wave and heat equations [3], which traces back to the elliptic case (see [11], [17]). We split $(u^\varepsilon\,,\,\tau^\varepsilon)$ into two parts. The first has a "correct energy" and a corrector result is obtained. The second tends weakly but not strongly to zero. It dissipates the energy gap in its effort to reconcile the energy of the initial conditions of $(u^\varepsilon, \tau^\varepsilon)$ to the energy of the initial conditions of the limit (u, τ).

Throughout this Section we assume that the convergences needed to obtain Theorem 3.1 take place for the whole sequence ε (see Section 3 and the introduction of Section 4).

Let $(u^\varepsilon, \tau^\varepsilon)$ be the solution of the system (1.1)-(1.7). Define $e^\varepsilon \in L^\infty(0,T)$ by

$$(5.1) \qquad \begin{cases} e^\varepsilon(t) = \dfrac{1}{2}\displaystyle\int_\Omega [\rho^\varepsilon \,|\frac{\partial u^\varepsilon}{\partial t}|^2 + A^\varepsilon \, grad\, u^\varepsilon \, grad\, u^\varepsilon + \beta^\varepsilon(\tau^\varepsilon)^2](x,t)\,dx \\ \qquad + \displaystyle\int_0^t \int_\Omega (K^\varepsilon \, grad\, \tau^\varepsilon \, grad\, \tau^\varepsilon)(x,s)\,dx\,ds; \end{cases}$$

we will improperly refer to e^ε as an "energy".

Recalling that equality (2.3) holds true we have

$$(5.2) \qquad e^\varepsilon(t) = E^\varepsilon + \int_0^t \int_\Omega f\,\frac{\partial u^\varepsilon}{\partial t}\,dx\,ds + \int_0^t < g\,, \tau^\varepsilon > ds \quad \text{in } (0,T)$$

where E^ε is the real number (the "energy of the initial conditions") defined by

$$(5.3) \qquad E^\varepsilon = \frac{1}{2}\int_\Omega [\rho^\varepsilon\,|b^\varepsilon|^2 + A^\varepsilon\, grad\, a^\varepsilon\, grad\, a^\varepsilon + \beta^\varepsilon\,|c^\varepsilon|^2]\,dx.$$

The solution (u, τ) of the homogenized system (3.1)-(3.7) satisfies an equality similar to (2.3). Defining $e^0 \in L^\infty(0,T)$ and $E^0 \in \mathbb{R}$ by

$$(5.4) \qquad \begin{cases} e^0(t) = \dfrac{1}{2}\displaystyle\int_\Omega [\bar\rho\,|\frac{\partial u}{\partial t}|^2 + A^0\, grad\, u\, grad\, u + (\bar\beta + \kappa)\,|\tau|^2](x,t)\,dx \\ \qquad + \displaystyle\int_0^t \int_\Omega (K^0\, grad\, \tau\, grad\, \tau)(x,s)\,dx\,ds \end{cases}$$

$$(5.5) \qquad E^0 = \frac{1}{2}\int_\Omega [\bar\rho\,|b^0|^2 + A^0\, grad\, a^0\, grad\, a^0 + (\bar\beta + \kappa)\,|c^0|^2]dx$$

we obtain

$$(5.6) \qquad e^0(t) = E^0 + \int_0^t \int_\Omega f\,\frac{\partial u}{\partial t}dx\,ds + \int_0^t < g\,, \tau > ds \quad \text{in } (0,T).$$

In view of the boundedness (2.4) and (1.12) of $(u^\varepsilon, \tau^\varepsilon)$ and of the initial conditions the sequences e^ε and E^ε are bounded in $L^\infty(0,T)$ and $\mathbb{R}$ respectively. We can thus extract a subsequence such that

$$(5.7) \qquad \begin{cases} e^{\varepsilon'} \rightharpoonup e \text{ weak} * \text{ in } L^\infty(0,T) \\ E^{\varepsilon'} \to E \text{ in } \mathbb{R} \end{cases}$$

(the first convergence actually takes places in the strong topology of $C^0(0,T)$ using Ascoli-Arzela's theorem and the analogue of (6.7) below), and of course we deduce from (5.2) that

$$e(t) = E + \int_0^t \int_\Omega f \frac{\partial u}{\partial t} \, dx \, ds + \int_0^t < g \, , \tau > ds \text{ in } (0,T). \tag{5.8}$$

The main difficulty resides in the lack of convergence of e^ε and E^ε to e^0 and E^0 ; in general

$$e(t) \neq e^0(t) \quad \text{and} \quad E \neq E^0; \tag{5.9}$$

see Remark 5.3 below for a detailed analysis of the behaviour of e^ε and E^ε.

This observation is at the root of the introduction of a solution $(\tilde{u}^\varepsilon \, , \tilde{\tau}^\varepsilon)$ corresponding to initial conditions $(\tilde{a}^\varepsilon \, , \tilde{b}^\varepsilon \, , \tilde{c}^\varepsilon)$ with a "correct energy".

We consider the initial conditions $(\tilde{a}^\varepsilon \, , \tilde{b}^\varepsilon \, , \tilde{c}^\varepsilon)$ defined by

$$\begin{cases} \tilde{c}^\varepsilon = c^0 \\ \tilde{b}^\varepsilon = b^0 \end{cases} \tag{5.10}$$

(see (3.17)-(3.20) for the definitions of b^0 and c^0)

$$\begin{cases} -\, div(A^\varepsilon \, grad \, \tilde{a}^\varepsilon - c^0(\gamma^\varepsilon - \gamma^0)) = -div(A^0 \, grad \, a^0) \text{ in } \Omega \\ \tilde{a}^\varepsilon \in H_0^1(\Omega) \end{cases} \tag{5.11}$$

where a^0 is the weak limit of a^ε (see (3.16)). Note that $\tilde{b}^\varepsilon$ and $\tilde{c}^\varepsilon$ actually do not depend on ε and that they are exactly the initial conditions $\partial u/\partial t \, (0)$ and $\tau(0)$ of the limit $(u \, , \tau)$ of $(u^\varepsilon \, , \tau^\varepsilon)$ (see Theorem 3.1).

We then define $(\tilde{u}^\varepsilon \, , \tilde{\tau}^\varepsilon)$ and $(v^\varepsilon \, , \theta^\varepsilon)$ as the solutions of the following systems.

$$\begin{cases} \rho^\varepsilon \dfrac{\partial^2 \tilde{u}^\varepsilon}{\partial t^2} - div(A^\varepsilon \, grad \, \tilde{u}^\varepsilon - \gamma^\varepsilon \, \tilde{\tau}^\varepsilon) = f \quad \text{in } \Omega \times (0,T) \\ \beta^\varepsilon \dfrac{\partial \tilde{\tau}^\varepsilon}{\partial t} - div(K^\varepsilon \, grad \, \tilde{\tau}^\varepsilon) + \gamma^\varepsilon \, grad \dfrac{\partial \tilde{u}^\varepsilon}{\partial t} = g \quad \text{in } \Omega \times (0,T) \\ \tilde{u}^\varepsilon(x,t) = 0 \quad \text{on } \partial\Omega \times (0,T) \\ \tilde{\tau}^\varepsilon(x,t) = 0 \quad \text{on } \partial\Omega \times (0,T) \\ \tilde{u}^\varepsilon(x,0) = \tilde{a}^\varepsilon(x) \quad \text{in } \Omega \\ \dfrac{\partial \tilde{u}^\varepsilon}{\partial t}(x,0) = \tilde{b}^\varepsilon(x) = b^0(x) \quad \text{in } \Omega \\ \tilde{\tau}^\varepsilon(x,0) = \tilde{c}^\varepsilon(x) = c^0(x) \quad \text{in } \Omega \end{cases} \tag{5.12}$$

$$(5.13)\qquad \begin{cases} \rho^\varepsilon \dfrac{\partial^2 v^\varepsilon}{\partial t^2} - div(A^\varepsilon\, grad\, v^\varepsilon - \gamma^\varepsilon\, \theta^\varepsilon) = 0 \quad \text{in } \Omega \times (0,T) \\ \beta^\varepsilon \dfrac{\partial \theta^\varepsilon}{\partial t} - div(K^\varepsilon\, grad\, \theta^\varepsilon) + \gamma^\varepsilon\, grad \dfrac{\partial v^\varepsilon}{\partial t} = 0 \quad \text{in } \Omega \times (0,T) \\ v^\varepsilon(x,t) = 0 \quad \text{on } \partial\Omega \times (0,T) \\ \tau^\varepsilon(x,t) = 0 \quad \text{on } \partial\Omega \times (0,T) \\ v^\varepsilon(x,0) = a^\varepsilon(x) - \tilde{a}^\varepsilon(x) \quad \text{in } \Omega \\ \dfrac{\partial v^\varepsilon}{\partial t}(x,0) = b^\varepsilon(x) - \tilde{b}^\varepsilon(x) = b^\varepsilon(x) - b^0(x) \quad \text{in } \Omega \\ \theta^\varepsilon(x,0) = c^\varepsilon(x) - \tilde{c}^\varepsilon(x) = c^\varepsilon(x) - c^0(x) \quad \text{in } \Omega. \end{cases}$$

Denote by $\tilde{e}^\varepsilon(t)$ and $\tilde{E}^\varepsilon$ the "energy" associated to $(\tilde{u}^\varepsilon, \tilde{\tau}^\varepsilon)$ and the "energy associated to the initial conditions" $(\tilde{a}^\varepsilon, \tilde{b}^\varepsilon, \tilde{c}^\varepsilon)$ respectively and by $\eta^\varepsilon(t)$ and H^ε the corresponding "energies" associated to $(v^\varepsilon, \theta^\varepsilon)$, namely

$$(5.14)\qquad \begin{cases} \tilde{e}^\varepsilon(t) = \dfrac{1}{2}\displaystyle\int_\Omega [\rho^\varepsilon\, |\frac{\partial \tilde{u}^\varepsilon}{\partial t}|^2 + A^\varepsilon\, grad\, \tilde{u}^\varepsilon\, grad\, \tilde{u}^\varepsilon + \beta^\varepsilon\, |\tilde{\tau}^\varepsilon|^2](x,t)\, dx \\ \qquad + \displaystyle\int_0^t \int_\Omega (K^\varepsilon\, grad\, \tilde{\tau}^\varepsilon\, grad\, \tilde{\tau}^\varepsilon)(x,s)\, dx\, ds \\ \tilde{E}^\varepsilon = \dfrac{1}{2}\displaystyle\int_\Omega [\rho^\varepsilon\, |b^0|^2 + A^\varepsilon\, grad\, \tilde{a}^\varepsilon\, grad\, \tilde{a}^\varepsilon + \beta^\varepsilon\, |c^0|^2]\, dx \end{cases}$$

$$(5.15)\qquad \begin{cases} \eta^\varepsilon(t) = \dfrac{1}{2}\displaystyle\int_\Omega [\rho^\varepsilon\, |\frac{\partial v^\varepsilon}{\partial t}|^2 + A^\varepsilon\, grad\, v^\varepsilon\, grad\, v^\varepsilon + \beta^\varepsilon\, |\theta^\varepsilon|^2](x,t)\, dx \\ \qquad + \displaystyle\int_0^t \int_\Omega (K^\varepsilon\, grad\, \theta^\varepsilon\, grad\, \theta^\varepsilon)(x,s)\, dx\, ds \\ H^\varepsilon = \dfrac{1}{2}\displaystyle\int_\Omega [\rho^\varepsilon\, |b^\varepsilon - b^0|^2 + A^\varepsilon\, grad(a^\varepsilon - \tilde{a}^\varepsilon)\, grad(a^\varepsilon - \tilde{a}^\varepsilon) \\ \qquad\qquad + \beta^\varepsilon\, |c^\varepsilon - c^0|^2](x)\, dx \end{cases}$$

Equalities similar to (5.2) (or (2.3)) hold for $(\tilde{u}^\varepsilon, \tilde{\tau}^\varepsilon)$ and $(v^\varepsilon, \theta^\varepsilon)$ and yield :

$$(5.16a)\qquad \tilde{e}^\varepsilon(t) = \tilde{E}^\varepsilon + \int_0^t \int_\Omega f \frac{\partial \tilde{u}^\varepsilon}{\partial t}\, dx\, ds + \int_0^t < g, \tilde{\tau}^\varepsilon > ds \quad \text{in } (0,T)$$

$$(5.16b)\qquad \eta^\varepsilon(t) = H^\varepsilon \quad \text{in } (0,T).$$

Obviously

$$(5.17)\qquad \begin{cases} u^\varepsilon = \tilde{u}^\varepsilon + v^\varepsilon \\ \tau^\varepsilon = \tilde{\tau}^\varepsilon + \tau^\varepsilon. \end{cases}$$

We will successively study $(\tilde{u}^\varepsilon, \tilde{\tau}^\varepsilon)$ and $(v^\varepsilon, \theta^\varepsilon)$. The obtained results will be summed up in Theorem 5.3 below, which describes in detail the structure of $(u^\varepsilon, \tau^\varepsilon)$.

Denote by $\tilde{s}^\varepsilon$ and σ^ε the entropies associated to $(\tilde{u}^\varepsilon, \tilde{\tau}^\varepsilon)$ and $(v^\varepsilon, \theta^\varepsilon)$, i.e.

$$(5.18a)\qquad \begin{cases} \tilde{s}^\varepsilon = \beta^\varepsilon \tilde{\tau}^\varepsilon + \gamma^\varepsilon \, grad\, \tilde{u}^\varepsilon \\ \sigma^\varepsilon = \beta^\varepsilon \theta^\varepsilon + \gamma^\varepsilon \, grad\, v^\varepsilon \end{cases}$$

and recall that the entropy associated to (u, τ) is

$$(5.18b)\qquad s = (\bar{\beta} + \kappa)\tau + \gamma^0 \, grad\, u$$

(see (4.19)). The initial conditions on $\tilde{s}^\varepsilon$, σ^ε and s are given by (see Proposition 2.1)

$$(5.19)\qquad \begin{cases} \tilde{s}^\varepsilon(0) = \tilde{d}^\varepsilon \\ \sigma^\varepsilon(0) = d^\varepsilon - \tilde{d}^\varepsilon \\ s(0) = d^0 \end{cases}$$

where $\tilde{d}^\varepsilon$ is defined by

$$(5.20)\qquad \tilde{d}^\varepsilon = \beta^\varepsilon \tilde{c}^\varepsilon + \gamma^\varepsilon \, grad\, \tilde{a}^\varepsilon = \beta^\varepsilon c^0 + \gamma^\varepsilon \, grad\, \tilde{a}^\varepsilon$$

and d^ε, d^0 are defined by (3.19a), (3.20).

THEOREM 5.1. *When ε tends to zero the following convergences hold true :*

$$(5.21)\qquad \begin{cases} \tilde{a}^\varepsilon \rightharpoonup a^0 \text{ weakly in } H_0^1(\Omega) \\ \tilde{d}^\varepsilon \rightharpoonup d^0 \text{ weakly in } L^2(\Omega) \end{cases}$$

$$(5.22)\qquad \begin{cases} \tilde{e}^\varepsilon \to e^0 \text{ strongly in } C^0(0,T) \\ \tilde{E}^\varepsilon \to E^0 \text{ in } \mathbb{R} \end{cases}$$

$$(5.23)\qquad \begin{cases} \tilde{u}^\varepsilon \rightharpoonup u \text{ weak } * \text{ in } L^\infty(0,T; H_0^1(\Omega)) \\ \dfrac{\partial \tilde{u}^\varepsilon}{\partial t} \rightharpoonup \dfrac{\partial u}{\partial t} \text{ weak } * \text{ in } L^\infty(0,T; L^2(\Omega)) \\ \tilde{\tau}^\varepsilon \rightharpoonup \tau \text{ weak } * \text{ in } L^\infty(0,T; L^2(\Omega)) \cap L^2(0,T; H_0^1(\Omega)) \end{cases}$$

where (u, τ) is the solution of the homogenized system (3.1)-(3.7).

Moreover the following corrector result hold true :

$$(5.24)\qquad \begin{cases} \dfrac{\partial \tilde{u}^\varepsilon}{\partial t} \to \dfrac{\partial u}{\partial t} \text{ strongly in } C^0(0,T; L^2(\Omega)) \\ \tilde{\tau}^\varepsilon \to \tau \text{ strongly in } C^0(0,T; L^2(\Omega)) \end{cases}$$

$$(5.25)\qquad \begin{cases} grad\,\tilde{u}^\varepsilon - P^\varepsilon\, grad\, u - \tau\, grad\, y^\varepsilon \;\to\; 0 \text{ strongly in } C^0(0,T;(L^1(\Omega))^N) \\ grad\,\tilde{\tau}^\varepsilon - Q^\varepsilon\, grad\,\tau \;\to\; 0 \text{ strongly in } L^2(0,T;(L^1(\Omega))^N) \end{cases}$$

where the corrector matrix P^ε (associated to A^ε) is defined by (3.10) and y^ε is defined by (3.13) ; Q^ε is the corrector matrix associated to K^ε. □

REMARK 5.1. When additional regularity holds for $(u\,,\,\tau)$, the strong convergences in (5.25) take place in better spaces than $C^0(0,T;(L^1(\Omega))^N)$ and $L^2(0,T;(L^1(\Omega))^N)$. For example when u and τ belong to $C^0(0,T;C^1(\bar{\Omega}))$ and $L^2(0,T;C^1(\bar{\Omega}))$ and not only to $C^0(0,T;H_0^1(\Omega))$ and $L^2(0,T;H_0^1(\Omega))$, one obtains the strong convergence in $C^0(0,T;(L^2(\Omega))^N)$ and $L^2(0,T;(L^2(\Omega))^N)$. The same result holds true if $grad\, y^\varepsilon$, P^ε and Q^ε are bounded in $(L^\infty(\Omega))^N$ and $(L^\infty(\Omega))^{N^2}$ respectively (and not only in L^2) while $(u\,,\,\tau)$ merely belong to the "natural" space defined by (2.5). □

REMARK 5.2. Note that the following convergence results hold for the initial conditions of the "quantities with tilde" :

$$(5.26)\qquad \begin{cases} \tilde{u}^\varepsilon(0) = \tilde{a}^\varepsilon \rightharpoonup a^0 = u(0) \text{ weakly in } H_0^1(\Omega) \\ \dfrac{\partial \tilde{u}^\varepsilon}{\partial t}(0) = b^0 = \dfrac{\partial u}{\partial t}(0) \\ (\rho^\varepsilon \dfrac{\partial \tilde{u}^\varepsilon}{\partial t})(0) = \rho^\varepsilon\, b^0 \rightharpoonup \bar{\rho}\, b^0 = (\bar{\rho}\dfrac{\partial u}{\partial t})(0) \text{ weakly in } L^2(\Omega) \\ \tilde{\tau}^\varepsilon(0) = c^0 = \tau(0) \\ \tilde{s}^\varepsilon(0) = \tilde{d}^\varepsilon \rightharpoonup d^0 = s(0) \text{ weakly in } L^2(\Omega). \end{cases}$$

Let us now turn to the study of the solution $(v^\varepsilon\,,\,\theta^\varepsilon)$ of (5.13).

THEOREM 5.2. *When ε tends to zero the following convergences hold true :*

$$(5.27)\qquad \begin{cases} v^\varepsilon \rightharpoonup 0 \text{ weak } * \text{ in } L^\infty(0,T;H_0^1(\Omega)) \\ \dfrac{\partial \tilde{v}^\varepsilon}{\partial t} \rightharpoonup 0 \text{ weak } * \text{ in } L^\infty(0,T;L^2(\Omega)) \\ \theta^\varepsilon \rightharpoonup 0 \text{ weak } * \text{ in } L^\infty(0,T;L^2(\Omega)) \cap L^2(0,T;H_0^1(\Omega)). \end{cases}$$

The convergences in (5.27) take place in the strong topologies if and only if the initial conditions $(a^\varepsilon\,,\,b^\varepsilon\,,\,c^\varepsilon)$ satisfy

$$(5.28)\qquad \begin{cases} a^\varepsilon - \tilde{a}^\varepsilon \to 0 \text{ strongly in } H_0^1(\Omega) \\ b^\varepsilon - b^0 \to 0 \text{ strongly in } L^2(\Omega) \\ c^\varepsilon - c^0 \to 0 \text{ strongly in } L^2(\Omega). \end{cases}$$

□

REMARK 5.3. We will also prove that (5.28) is equivalent to

$$(5.29)\qquad E^\varepsilon \to E^0 \text{ in } \mathbb{R}$$

whereas in general

$$\liminf_{\varepsilon \to 0} E^\varepsilon \geq E^0. \tag{5.30}$$

Convergences (5.28) will also be proved to be equivalent to

$$e^\varepsilon \to e^0 \text{ strongly in } C^0(0,T) \tag{5.31}$$

as well as to

$$H^\varepsilon \to 0. \tag{5.32}$$

This remark complements assertion (5.9). □

REMARK 5.4. Observe that in view of (5.24)

$$\begin{cases} \tilde{\tau}^\varepsilon \to \tau \text{ strongly in } C^0(0,T;L^2(\Omega)) \\ \dfrac{\partial \tilde{u}^\varepsilon}{\partial t} \to \dfrac{\partial u}{\partial t}\; C^0(0,T;L^2(\Omega)) \end{cases} \tag{5.33}$$

but that only weak convergence results hold for θ^ε and $\partial v^\varepsilon/\partial t$: they do not converge strongly in $L^\infty(0,T;L^2(\Omega))$ except if (5.28) holds true. Actually we believe that in general

$$\begin{cases} \theta^\varepsilon \not\to 0 \quad \text{a.e. in } \Omega \times (0,T) \\ \dfrac{\partial v^\varepsilon}{\partial t} \not\to 0 \quad \text{a.e. in } \Omega \times (0,T); \end{cases} \tag{5.34}$$

(see [4], [6] for a supporting analysis of this belief based on a formal asymptotic expansion). □

The results obtained in Theorems 5.1 and 5.2 are merged in the following theorem that provides a detailed picture of the structure of $(u^\varepsilon, \tau^\varepsilon)$ itself.

THEOREM 5.3. *When ε tends to zero, the solution $(u^\varepsilon, \tau^\varepsilon)$ of (1.1)-(1.7) has the following structure :*

$$\begin{cases} \dfrac{\partial u^\varepsilon}{\partial t} = \dfrac{\partial u}{\partial t} + \dfrac{\partial v^\varepsilon}{\partial t} + r_1^\varepsilon \\ \tau^\varepsilon = \tau + \theta^\varepsilon + r_2^\varepsilon \end{cases} \tag{5.35}$$

$$\begin{cases} grad\, u^\varepsilon = P^\varepsilon\, grad\, u + \tau\, grad\, y^\varepsilon + grad\, v^\varepsilon + R_1^\varepsilon \\ grad\, \tau^\varepsilon = Q^\varepsilon\, grad\, \tau + grad\, \theta^\varepsilon + R_2^\varepsilon \end{cases} \tag{5.36}$$

where $(u\,,\,\tau)$ is the solution of the homogenized system (3.1)-(3.7), P^ε and Q^ε are the corrector matrices associated to A^ε and K^ε respectively (see (3.10)) and y^ε is defined by (3.13) ; $(v^\varepsilon\,,\,\theta^\varepsilon)$ is the solution of (5.13) and satisfies

$$\text{(5.37)}\qquad \begin{cases} \dfrac{\partial v^\varepsilon}{\partial t}\,,\,\theta^\varepsilon \rightharpoonup 0 \text{ weak } * \text{ in } L^\infty(0,T;L^2(\Omega)) \\ grad\, v^\varepsilon \rightharpoonup 0 \text{ weak } * \text{ in } L^\infty(0,T;(L^2(\Omega))^N) \\ grad\, \theta^\varepsilon \rightharpoonup 0 \text{ weakly in } L^2(0,T;(L^2(\Omega))^N) \end{cases}$$

while

$$\text{(5.38)}\qquad r_1^\varepsilon\,,\,r_2^\varepsilon \to 0 \text{ strongly in } C^0(0,T;L^2(\Omega))$$

$$\text{(5.39)}\qquad \begin{cases} R_1^\varepsilon \to 0 \text{ strongly in } C^0(0,T;(L^1(\Omega))^N) \\ R_2^\varepsilon \to 0 \text{ strongly in } L^2(0,T;(L^1(\Omega))^N). \end{cases}$$

The strong convergences in (5.39) take place in better spaces when additional regularity assumptions hold (see Remark 5.1). Moreover the convergences in (5.37) are strong (and thus the terms $\partial v^\varepsilon/\partial t\,,\,\theta^\varepsilon\,,\,grad\, v^\varepsilon\,,\,grad\,\theta^\varepsilon$ can be removed in (5.35), (5.36)) if and only if (5.28) holds true.

REMARK 5.5. In the case of the uncoupled wave and heat equations, which corresponds to the case $\gamma^\varepsilon \equiv 0$ in (1.1), (1.2) the behaviour of each of the four terms of the "energy" $\eta^\varepsilon(t) = H^\varepsilon$ associated to $(v^\varepsilon\,,\,\theta^\varepsilon)$ can be more precisely analyzed. In this case there are two uncoupled energies

(5.40)

$$\begin{cases} \dfrac{1}{2}\displaystyle\int_\Omega [\rho^\varepsilon\,|\frac{\partial v^\varepsilon}{\partial t}|^2 + A^\varepsilon\, grad\, v^\varepsilon\, grad\, v^\varepsilon](x,t)\,dx = H_w^\varepsilon \in \mathbb{R} \\ \dfrac{1}{2}\displaystyle\int_\Omega (\beta^\varepsilon\,|\theta^\varepsilon|^2)(x,t)dx + \int_0^t\int_\Omega (K^\varepsilon\, grad\,\theta^\varepsilon\, grad\,\theta^\varepsilon)(x,s)dx\,ds = H_h^\varepsilon \in \mathbb{R}. \end{cases}$$

Extracting a subsequence such that

$$H_w^{\varepsilon'} \to H_w \text{ in } \mathbb{R} \quad , \quad H_h^{\varepsilon'} \to H_h \text{ in } \mathbb{R}$$

we prove in [3] that

$$\text{(5.41)}\qquad \begin{cases} \dfrac{1}{2}\displaystyle\int_\Omega (\rho^\varepsilon\,|\frac{\partial v^\varepsilon}{\partial t}|^2)(x,t)dx \rightharpoonup \frac{1}{2}H_w \text{ weak } * \text{ in } L^\infty(0,T) \\ \dfrac{1}{2}\displaystyle\int_\Omega (A^\varepsilon\, grad\, v^\varepsilon\, grad\, v^\varepsilon)(x,t)dx \rightharpoonup \frac{1}{2}H_w \text{ weak } * \text{ in } L^\infty(0,T) \end{cases}$$

$$\text{(5.42)}\qquad \begin{cases} \dfrac{1}{2}\displaystyle\int_\Omega (\beta^\varepsilon\,|\theta^\varepsilon|^2)(x,t)dx \rightharpoonup 0 \\ \qquad \text{weak } * \text{ in } L^\infty(0,T) \text{ and strongly in } C^0(\delta,T) \text{ for any } \delta > 0 \\ \displaystyle\int_0^t\int_\Omega (K^\varepsilon\, grad\,\theta^\varepsilon\, grad\,\theta^\varepsilon)(x,s)dx\,ds \rightharpoonup H_h \\ \qquad \text{weak } * \text{ in } L^\infty(0,T) \text{ and strongly in } C^0(\delta,T) \text{ for any } \delta > 0. \end{cases}$$

Thus θ^ε is concentrated in an initial layer about the time $t = 0$ while the energy of $\partial v^\varepsilon/\partial t$ and $grad\, v^\varepsilon$ permeates all times with equipartition of the energy H_w.

We were not able to carry over such a precise analysis for the case of the coupled thermoelasticity system (1.1)-(1.7). In particular we were unable to prove that the limits of the functions $t \to \frac{1}{2}\int_\Omega \rho^\varepsilon\, |\partial v^\varepsilon/\partial t|^2\, dx$, $t \to \frac{1}{2}\int_\Omega A^\varepsilon\, grad\, v^\varepsilon\, grad\, v^\varepsilon\, dx$ as well as of $t \to \frac{1}{2}\int_\Omega (\beta^\varepsilon\, |\theta^\varepsilon|^2)\, dx$ are in general different from 0 for any positive time, although we believe that such is the case. This opinion is substantiated by the formal asymptotics and the numerical computations presented in [4], [6], which suggest the presence of fast time oscillations of θ^ε and $\partial v^\varepsilon/\partial t$. □

6. PROOFS OF THE CORRECTOR RESULT.

PROOF OF THEOREM 5.1.

FIRST PART : PROOF OF (5.21)-(5.23). As in (4.3), (4.4) two applications of the div-curl lemma prove that the solution $\tilde{a}^\varepsilon$ of (5.11) satisfies

$$\tilde{a}^\varepsilon \rightharpoonup a^0 \text{ weakly in } H_0^1(\Omega). \tag{6.1}$$

In view of equation (5.11) and of Proposition 4.1,

$$\begin{cases} \gamma^\varepsilon\, grad\, \tilde{a}^\varepsilon = \gamma^\varepsilon [P^\varepsilon\, grad\, a^0 + c^0\, grad\, y^\varepsilon + r^\varepsilon] \\ \qquad \rightharpoonup \gamma^0\, grad\, a^0 + c^0\, \kappa \\ \qquad \text{in } \mathcal{D}'(\Omega) \text{and weakly in } L^2(\Omega) \end{cases} \tag{6.2}$$

since $\gamma^\varepsilon\, grad\, \tilde{a}^\varepsilon$ is bounded in $L^2(\Omega)$. Thus

$$\tilde{d}^\varepsilon = \beta^\varepsilon\, c^0 + \gamma^\varepsilon\, grad\, \tilde{a}^\varepsilon \rightharpoonup (\bar{\beta} + \kappa) c^0 + \gamma^0\, grad\, a^0 = d^0 \text{weakly in } L^2(\Omega) \tag{6.3}$$

and (5.21) is proved.

In view of (5.21) and of $\tilde{b}^\varepsilon = b^0$, Theorem 3.1 immediately implies (5.23), where $(u\, ,\, \tau)$ is the solution of the homogenized system (3.1)-(3.7).

Let us now prove (5.22). Obviously we have

$$\begin{cases} \displaystyle\int_\Omega \rho^\varepsilon\, |\tilde{b}^\varepsilon|^2\, dx = \int_\Omega \rho^\varepsilon\, |b^0|^2\, dx \to \int_\Omega \bar{\rho}\, |b^0|^2\, dx \\ \displaystyle\int_\Omega \beta^\varepsilon\, |\tilde{c}^\varepsilon|^2\, dx = \int_\Omega \beta^\varepsilon\, |c^0|^2\, dx \to \int_\Omega \bar{\beta}\, |c^0|^2\, dx. \end{cases} \tag{6.4}$$

On the other hand, multiplication of (5.11) by $\tilde{a}^\varepsilon$, integration by parts and use of (6.3) and (6.1) imply that

(6.5)

$$\begin{cases} \displaystyle\int_\Omega A^\varepsilon\, grad\, \tilde{a}^\varepsilon\, grad\, \tilde{a}^\varepsilon\, dx \\ \displaystyle = \int_\Omega c^0\, \gamma^\varepsilon\, grad\, \tilde{a}^\varepsilon\, dx - \int_\Omega c^0\, \gamma^0\, grad\, \tilde{a}^\varepsilon\, dx + \int_\Omega A^0\, grad\, a^0\, grad\, \tilde{a}^\varepsilon\, dx \\ \displaystyle \to \int_\Omega c^0 (\gamma^0\, grad\, a^0 + c^0\, \kappa) dx - \int_\Omega c^0\, \gamma^0\, grad\, a^0\, dx + \int_\Omega A^0\, grad\, a^0\, grad\, a^0\, dx \\ \displaystyle = \int_\Omega A^0\, grad\, a^0\, grad\, a^0\, dx + \int_\Omega \kappa\, |c^0|^2\, dx. \end{cases}$$

By virtue of the definitions (5.14) and (5.5) we proved that $\tilde{E}^\varepsilon$ tends to E^0.

In view of (5.16a) the first part of (5.22) follows from

$$(6.6)\quad \begin{cases} \int_0^t \int_\Omega f \frac{\partial \tilde{u}^\varepsilon}{\partial t} \, dx\, ds + \int_0^t < g\,,\, \tilde{\tau}^\varepsilon > ds \\ \to \int_0^t \int_\Omega f \frac{\partial u}{\partial t} \, dx\, ds + \int_0^t < g\,,\, \tau > ds \text{ strongly in } C^0(0,T). \end{cases}$$

Because of (5.23) this convergence is clear in $L^\infty(0,T)$ weak star. The strong convergence in $C^0(0,T)$ then follows from Ascoli-Arzela's theorem and from

$$(6.7)\quad \begin{cases} \left| \int_t^{t+h} \int_\Omega f \frac{\partial \tilde{u}^\varepsilon}{\partial t} \, dx\, ds \right| \leq \left\| \frac{\partial \tilde{u}^\varepsilon}{\partial t} \right\|_{L^2(0,T;L^2(\Omega))} \|f\|_{L^2(t,t+h;L^2(\Omega))} \\ \left| \int_t^{t+h} < g\,,\, \tilde{\tau}^\varepsilon > ds \right| \leq \|\tilde{\tau}^\varepsilon\|_{L^2(0,T;H_0^1(\Omega))} \|g\|_{L^2(t,t+h;H^{-1}(\Omega))}. \end{cases}$$

SECOND PART : THE HEART OF THE MATTER. In a manner similar to that of the elliptic case ([11][17] ; see also the proof of Proposition 4.1 above), and of the wave and heat equations ([3]) we consider for given ϕ and ψ in $C^\infty(0,T;\mathcal{D}(\Omega))$ the function $\tilde{X}^\varepsilon$ defined by

(6.8)

$$\begin{cases} \tilde{X}^\varepsilon(t) = \frac{1}{2} \int_\Omega [\rho^\varepsilon \left| \frac{\partial \tilde{u}^\varepsilon}{\partial t} - \frac{\partial \phi}{\partial t} \right|^2 + A^\varepsilon(grad\, \tilde{u}^\varepsilon - P^\varepsilon\, grad\, \phi - \psi\, grad\, y^\varepsilon) \\ \qquad (grad\, \tilde{u}^\varepsilon - P^\varepsilon\, grad\, \phi - \psi\, grad\, y^\varepsilon) + \beta^\varepsilon\, |\tilde{\tau}^\varepsilon - \psi|^2](x,t) dx \\ + \int_0^t \int_\Omega (K^\varepsilon(grad\, \tilde{\tau}^\varepsilon - Q^\varepsilon\, grad\, \psi)(grad\, \tilde{\tau}^\varepsilon - Q^\varepsilon\, grad\, \psi))(x,s) dx\, ds \\ = I^\varepsilon(t) + II^\varepsilon(t) + III^\varepsilon(t) \end{cases}$$

where

$$\begin{aligned} I^\varepsilon(t) =& \frac{1}{2} \int_\Omega [\rho^\varepsilon \left| \frac{\partial \tilde{u}^\varepsilon}{\partial t} \right|^2 + A^\varepsilon\, grad\, \tilde{u}^\varepsilon\, grad\, \tilde{u}^\varepsilon + \beta^\varepsilon\, |\tilde{\tau}^\varepsilon|^2]\, dx \\ &+ \int_0^t \int_\Omega K^\varepsilon\, grad\, \tilde{\tau}^\varepsilon\, grad\, \tilde{\tau}^\varepsilon\, dx\, ds \end{aligned}$$

$$\begin{aligned} II^\varepsilon(t) =& - \int_\Omega [\rho^\varepsilon \frac{\partial \tilde{u}^\varepsilon}{\partial t} \frac{\partial \phi}{\partial t} + A^\varepsilon(P^\varepsilon\, grad\, \phi + \psi\, grad\, y^\varepsilon)\, grad\, \tilde{u}^\varepsilon + \beta^\varepsilon\, \tilde{\tau}^\varepsilon\, \psi] dx \\ &- 2 \int_0^t \int_\Omega K^\varepsilon\, Q^\varepsilon\, grad\, \psi\, grad\, \tilde{\tau}^\varepsilon\, dx\, ds \end{aligned}$$

$$\begin{aligned} &III^\varepsilon(t) = \\ &= \frac{1}{2} \int_\Omega [\rho^\varepsilon \left| \frac{\partial \phi}{\partial t} \right|^2 + A^\varepsilon(P^\varepsilon\, grad\, \phi + \psi\, grad\, y^\varepsilon)(P^\varepsilon\, grad\, \phi + \psi\, grad\, y^\varepsilon) + \beta^\varepsilon\, |\psi|^2] dx \\ &+ \int_0^t \int_\Omega K^\varepsilon\, Q^\varepsilon\, grad\, \psi\, Q^\varepsilon\, grad\, \psi\, dx\, ds. \end{aligned}$$

We will successively pass to the limit in $I^\varepsilon, III^\varepsilon$ and II^ε. Since I^ε is precisely $\tilde{e}^\varepsilon$ defined by (5.14), (5.22) and definition (5.4) of e^0 yield

$$
(6.9) \quad \begin{cases} I^\varepsilon \to \dfrac{1}{2} \displaystyle\int_\Omega [\bar{\rho} |\frac{\partial u}{\partial t}|^2 + A^0 \, grad\, u \, grad\, u + (\bar{\beta} + \kappa)|\tau|^2](x,t) dx \\ \quad + \displaystyle\int_0^t \int_\Omega (K^0 \, grad\, \tau \, grad\, \tau)(x,s) dx\, ds \quad \text{strongly in } C^0(0,T). \end{cases}
$$

Note that III^ε and its time derivative $\partial III^\varepsilon / \partial t$ are bounded in $L^\infty(0,T)$. Thus III^ε converges strongly in $C^0(0,T)$ to some limit which can be identified in the sense of distributions using a test function $\varphi \in \mathcal{D}(0,T)$. Consider for example the term

$$
\begin{aligned}
&\int_0^T \varphi\, dt \int_\Omega A^\varepsilon (P^\varepsilon \, grad\, \phi + \psi \, grad\, y^\varepsilon)(P^\varepsilon \, grad\, \phi + \psi \, grad\, y^\varepsilon) dx \\
&= \int_0^T \int_\Omega \varphi (\sum_i A^\varepsilon P^\varepsilon e_i \frac{\partial \phi}{\partial x_i})(\sum_j P^\varepsilon e_j \frac{\partial \phi}{\partial x_j} + \psi \, grad\, y^\varepsilon) dx\, dt \\
&+ \int_0^T \int_\Omega \varphi \psi (A^\varepsilon \, grad\, y^\varepsilon - (\gamma^\varepsilon - \gamma^0))(\sum_j P^\varepsilon e_j \frac{\partial \phi}{\partial x_j} + \psi \, grad\, y^\varepsilon) dx\, dt \\
&+ \int_0^T \int_\Omega \varphi \psi (\gamma^\varepsilon - \gamma^0)(P^\varepsilon \, grad\, \phi + \psi \, grad\, y^\varepsilon) dx\, dt.
\end{aligned}
$$

Because the divergences in x of $A^\varepsilon P^\varepsilon e_i$ and of $(A^\varepsilon \, grad\, y^\varepsilon - (\gamma^\varepsilon - \gamma^0))$ as well as the curl in x and the time derivative of $P^\varepsilon e_j$ and of $grad\, y^\varepsilon$ are compact in $H^{-1}(\Omega \times (0,T))$, a time dependent version of the div-curl lemma (see [12], [18], or [3] for a precise statement) together with the definitions (3.11) and (3.14) of γ^0 and κ permit to obtain in the limit

$$
\begin{aligned}
&\int_0^T \int_\Omega \varphi (A^0 \, grad\, \phi)(grad\, \phi + 0) dx\, dt \\
&+ \int_0^T \int_\Omega \varphi \psi (0)(grad\, \phi + 0) dx\, dt \\
&+ \int_0^T \int_\Omega \varphi \psi [(\gamma^0 - \gamma^0) grad\, \phi + \psi (\kappa - \gamma^0\, 0)] dx\, dt.
\end{aligned}
$$

Thus

$$
(6.10) \quad \begin{cases} III^\varepsilon \to \dfrac{1}{2} \displaystyle\int_\Omega [\bar{\rho} |\frac{\partial \phi}{\partial t}|^2 + A^0 \, grad\, \phi \, grad\, \phi + (\bar{\beta} + \kappa)|\psi|^2](x,t) dx \\ \quad + \displaystyle\int_0^t \int_\Omega (K^0 \, grad\, \psi \, grad\, \psi)(x,s) dx\, ds \quad \text{strongly in } C^0(0,T). \end{cases}
$$

The second term II^ε is clearly bounded in $L^\infty(0,T)$. Writing II^ε as

$$(6.11)\quad \begin{cases} II^\varepsilon = -\displaystyle\int_\Omega [\rho^\varepsilon \frac{\partial \tilde{u}^\varepsilon}{\partial t}\frac{\partial \phi}{\partial t} + A^\varepsilon P^\varepsilon \, grad\, \phi \, grad\, \tilde{u}^\varepsilon \\ \qquad + \psi \, grad\, \tilde{u}^\varepsilon (A^\varepsilon \, grad\, y^\varepsilon - (\gamma^\varepsilon - \gamma^0)) \\ \qquad + \psi(\beta^\varepsilon \tilde{\tau}^\varepsilon + \gamma^\varepsilon \, grad\, \tilde{u}^\varepsilon) - \psi\gamma^0 \, grad\, \tilde{u}^\varepsilon] dx \\ \qquad - 2\displaystyle\int_0^t \int_\Omega K^\varepsilon Q^\varepsilon \, grad\, \psi \, grad\, \tilde{\tau}^\varepsilon \, dx\, ds \end{cases}$$

and using the time dependent version of the div-curl lemma as for III^ε, one can pass to the limit in the sense of distributions on $(0,T)$ in II^ε. It will be proved in the fourth part of this proof that each term actually converges strongly in $C^0(0,T)$ and thus that

$$(6.12)\quad \begin{cases} II^\varepsilon \to -\displaystyle\int_\Omega [\bar{\rho}\frac{\partial u}{\partial t}\frac{\partial \phi}{\partial t} + A^0 \, grad\, \phi \, grad\, u \\ \qquad + \psi \, grad\, u\, 0 + \psi((\bar{\beta} + \kappa)\tau + \gamma^0 \, grad\, u) - \psi\gamma^0 \, grad\, u] dx \\ \qquad - 2\displaystyle\int_0^t \int_\Omega K^0 \, grad\, \psi \, grad\, \tau \, dx\, ds \\ = -\displaystyle\int_\Omega [\bar{\rho}\frac{\partial u}{\partial t}\frac{\partial \phi}{\partial t} + A^0 \, grad\, \phi \, grad\, u + (\bar{\beta} + \kappa)\tau\psi] dx \\ \qquad - 2\displaystyle\int_0^t \int_\Omega K^0 \, grad\, \psi \, grad\, \tau \, dx\, ds \quad \text{strongly in } C^0(0,T). \end{cases}$$

From (6.8)-(6.12) we conclude that

(6.13)
$$\begin{cases} \tilde{X}^\varepsilon \to \dfrac{1}{2}\displaystyle\int_\Omega [\bar{\rho}|\frac{\partial u}{\partial t} - \frac{\partial \phi}{\partial t}|^2 + A^0 (grad\, u - grad\, \phi)(grad\, u - grad\, \phi) \\ \qquad + (\bar{\beta} + \kappa)|\tau - \psi|^2] dx \\ \qquad + \displaystyle\int_0^t \int_\Omega K^0 (grad\, \tau - grad\, \psi)(grad\, \tau - grad\, \psi) dx\, ds \quad \text{in } C^0(0,T) \end{cases}$$

which implies

(6.14)
$$\begin{cases} \limsup_{\varepsilon\to 0} \{\lambda_1 \|\dfrac{\partial \tilde{u}^\varepsilon}{\partial t} - \dfrac{\partial \phi}{\partial t}\|^2_{C^0(0,T;L^2(\Omega))} \\ \qquad + \lambda_1 \|grad\, \tilde{u}^\varepsilon - P^\varepsilon \, grad\, \phi - \psi \, grad\, y^\varepsilon\|^2_{C^0(0,T;(L^2(\Omega))^N)} \\ \qquad + \lambda_1 \|\tilde{\tau}^\varepsilon - \psi\|^2_{C^0(0,T;L^2(\Omega))} + 2\lambda_1 \|grad\, \tilde{\tau}^\varepsilon - Q^\varepsilon \, grad\, \psi\|^2_{(L^2(0,T;L^2(\Omega))^N)}\} \\ \le 2\{\lambda_2 \|\dfrac{\partial u}{\partial t} - \dfrac{\partial \phi}{\partial t}\|^2_{C^0(0,T;L^2(\Omega))} + \lambda_2 \|grad\, u - grad\, \phi\|^2_{C^0(0,T;(L^2(\Omega))^N)} \\ \qquad + (\lambda_2 + \|\kappa\|_{L^\infty(\Omega)}) \|\tau - \psi\|^2_{C^0(0,T;L^2(\Omega))} \\ \qquad + 2\lambda_2 \|grad\, \tau - grad\, \psi\|^2_{L^2(0,T;(L^2(\Omega))^N)}\}. \end{cases}$$

THIRD PART : END OF THE PROOF OF (5.24), (5.25). The corrector results (5.24) (5.25) for $(\tilde{u}^\varepsilon, \tilde{\tau}^\varepsilon)$ are then deduced from (6.14) with the help of the following decomposition :

(6.15)
$$\begin{cases} \dfrac{\partial \tilde{u}^\varepsilon}{\partial t} - \dfrac{\partial u}{\partial t} = \dfrac{\partial \tilde{u}^\varepsilon}{\partial t} - \dfrac{\partial \phi}{\partial t} + \dfrac{\partial \phi}{\partial t} - \dfrac{\partial u}{\partial t} \\ grad\, \tilde{u}^\varepsilon - P^\varepsilon\, grad\, u - \tau\, grad\, y^\varepsilon \\ = grad\, \tilde{u}^\varepsilon - P^\varepsilon\, grad\, \phi - \psi\, grad\, y^\varepsilon + P^\varepsilon(grad\, \phi - grad\, u) + (\psi - \tau) grad\, y^\varepsilon \end{cases}$$

(and the analogous formulas for $\tilde{\tau}^\varepsilon$). We choose ϕ and ψ in $C^\infty(0,T;\mathcal{D}(\Omega))$ such that $((\partial u/\partial t) - (\partial \phi/\partial t))$ and $(grad\, u - grad\, \phi)$ are small in $C^0(0,T;L^2(\Omega))$ and $C^0(0,T;(L^2(\Omega))^N)$ respectively, while $\tau - \psi$ is small in $C^0(0,T;L^2(\Omega)) \cap L^2(0,T;H_0^1(\Omega))$: this choice is possible because u belongs to $C^0(0,T;H_0^1(\Omega))$ with $\partial u/\partial t$ in $C^0(0,T;L^2(\Omega))$ and τ belongs to $C^0(0,T;L^2(\Omega)) \cap L^2(0,T;H_0^1(\Omega))$; recall also that P^ε and $grad\, y^\varepsilon$ are bounded in $(L^2(\Omega))^{N^2}$ and $(L^2(\Omega))^N$ respectively. This proves (5.24), (5.25). The "convergences in better spaces" announced in Remark 5.1 are proved in the same manner.

FOURTH PART : PROOF OF THE STRONG CONVERGENCE IN $C^0(0,T)$ OF THE TERMS OF II^ε (see (6.11), (6.12)). Using equation (5.12) leads to

$$(6.16) \quad \begin{cases} \dfrac{d}{dt} \displaystyle\int_\Omega \rho^\varepsilon \frac{\partial \tilde{u}^\varepsilon}{\partial t} \frac{\partial \phi}{\partial t}\, dx \\ = \displaystyle\int_\Omega \rho^\varepsilon \frac{\partial \tilde{u}^\varepsilon}{\partial t} \frac{\partial^2 \phi}{\partial t^2}\, dx + < \rho^\varepsilon \frac{\partial^2 \tilde{u}^\varepsilon}{\partial t^2}, \frac{\partial \phi}{\partial t} > \\ = \displaystyle\int_\Omega \rho^\varepsilon \frac{\partial \tilde{u}^\varepsilon}{\partial t} \frac{\partial^2 \phi}{\partial t^2}\, dx + \int_\Omega f \frac{\partial \phi}{\partial t}\, dx \\ \quad - \displaystyle\int_\Omega (A^\varepsilon\, grad\, \tilde{u}^\varepsilon - \gamma^\varepsilon\, \tilde{\tau}^\varepsilon) grad \frac{\partial \phi}{\partial t}\, dx. \end{cases}$$

By application of the Cauchy-Schwartz inequality this term is bounded in $L^2(0,T)$. Thus $t \to \int_\Omega \rho^\varepsilon (\partial \tilde{u}^\varepsilon/\partial t)(\partial \phi/\partial t)\, dx$ converges strongly in $C^0(0,T)$.

In a similar manner, (5.12) leads to

(6.17)
$$\begin{cases} \dfrac{d}{dt} \displaystyle\int_\Omega \psi(\beta^\varepsilon\, \tilde{\tau}^\varepsilon + \gamma^\varepsilon\, grad\, \tilde{u}^\varepsilon) dx = \\ = \displaystyle\int_\Omega \frac{\partial \psi}{\partial t} (\beta^\varepsilon\, \tilde{\tau}^\varepsilon + \gamma^\varepsilon\, grad\, \tilde{u}^\varepsilon) dx + \int_\Omega \psi \frac{\partial}{\partial t} (\beta^\varepsilon\, \tilde{\tau}^\varepsilon + \gamma^\varepsilon\, grad\, \tilde{u}^\varepsilon) dx \\ = \displaystyle\int_\Omega \frac{\partial \psi}{\partial t} (\beta^\varepsilon\, \tilde{\tau}^\varepsilon + \gamma^\varepsilon\, grad\, \tilde{u}^\varepsilon) dx - \int_\Omega K^\varepsilon\, grad\, \tilde{\tau}^\varepsilon\, grad\, \psi\, dx + < g, \psi > . \end{cases}$$

By application of the Cauchy-Schwartz inequality this term is bounded in $L^2(0,T)$. Thus $t \to \int_\Omega \psi(\beta^\varepsilon\, \tilde{\tau}^\varepsilon + \gamma^\varepsilon\, grad\, \tilde{u}^\varepsilon) dx$ converges strongly in $C^0(0,T)$.

Using Ascoli-Arzela's theorem, it is easy to see that the strong convergence in $C^0(0,T)$ of the term $\int_0^t \int_\Omega K^\varepsilon\, Q^\varepsilon\, grad\,\psi\; grad\,\tilde{\tau}^\varepsilon\, dx\, ds$ results from

(6.18)
$$\begin{cases} |\displaystyle\int_t^{t+h} \int_\Omega K^\varepsilon\, grad\,\tilde{\tau}^\varepsilon\, Q^\varepsilon\, grad\,\psi\, dx\, ds| \\ \le \|K^\varepsilon\, grad\,\tilde{\tau}^\varepsilon\|^2_{L^2(0,T;(L^2(\Omega))^N)}\, \|Q^\varepsilon\|_{(L^2(\Omega))^{N^2}}\, \|grad\,\psi\|_{L^2(t,t+h;(L^\infty(\Omega))^N)}. \end{cases}$$

Using integration by parts and the definition (3.10) of P^ε yields

(6.19a)
$$\begin{cases} \displaystyle\int_\Omega A^\varepsilon\, P^\varepsilon\, grad\,\phi\, grad\,\tilde{u}^\varepsilon\, dx = \sum_i \int_\Omega A^\varepsilon\, grad\,w_i^\varepsilon\, \frac{\partial\phi}{\partial x_i}\, grad\,\tilde{u}^\varepsilon\, dx \\ \displaystyle = -\sum_i < div(A^\varepsilon\, grad\,w_i^\varepsilon)\, \frac{\partial\phi}{\partial x_i}\,,\, \tilde{u}^\varepsilon > - \sum_i \int_\Omega A^\varepsilon\, grad\,w_i^\varepsilon\, grad\,\frac{\partial\phi}{\partial x_i}\, \tilde{u}^\varepsilon\, dx \\ \displaystyle = -\sum_i < div(A^0\, e_i)\, \frac{\partial\phi}{\partial x_i}\,,\, \tilde{u}^\varepsilon > - \sum_i \int_\Omega A^\varepsilon\, grad\,w_i^\varepsilon\, grad\,\frac{\partial\phi}{\partial x_i}\, \tilde{u}^\varepsilon\, dx \end{cases}$$

and similarly in view of the definition (3.13) of y^ε

(6.19b)
$$\begin{cases} \displaystyle\int_\Omega \psi(A^\varepsilon\, grad\,y^\varepsilon - (\gamma^\varepsilon - \gamma^0))grad\,\tilde{u}^\varepsilon\, dx \\ \displaystyle = -\int_\Omega grad\,\psi(A^\varepsilon\, grad\,y^\varepsilon - (\gamma^\varepsilon - \gamma^0))\tilde{u}^\varepsilon\, dx. \end{cases}$$

Thus defining k^ε by

(6.20)
$$\begin{cases} \displaystyle k^\varepsilon = [-\sum_i div(A^0\, e_i)\, \frac{\partial\phi}{\partial x_i} + div(\psi\,\gamma^0)] \\ \displaystyle \quad - \sum_i A^\varepsilon\, grad\,w_i^\varepsilon\, grad\,\frac{\partial\phi}{\partial x_i} - grad\,\psi(A^\varepsilon\, grad\,y^\varepsilon - (\gamma^\varepsilon - \gamma^0)), \end{cases}$$

the part $F^\varepsilon(t)$ of $II^\varepsilon(t)$ that has yet to be proved to converge strongly in $C^0(0,T)$ reads as

(6.21)
$$F^\varepsilon(t) = < k^\varepsilon\,,\, \tilde{u}^\varepsilon > (t).$$

But k^ε is the sum of a first term which is fixed in $C^0(0,T;H^{-1}(\Omega))$ and of a sequence which is bounded in $C^1(0,T;L^2(\Omega))$. Thus (for a subsequence)

(6.22)
$$k^\varepsilon \to k \text{ strongly in } C^0(0,T;H^{-1}(\Omega))$$

where k belongs to $C^0(0,T;H^{-1}(\Omega))$.

Consider h in $C^0(0,T;L^2(\Omega))$ and write for any $t \in (0,T)$

$$\begin{aligned} &F^\varepsilon(t) - < k\,,\, u > (t) = (< k^\varepsilon\,,\, \tilde{u}^\varepsilon > - < k\,,\, u >)(t) \\ &= (< k^\varepsilon - k\,,\, \tilde{u}^\varepsilon > + < k - h\,,\, \tilde{u}^\varepsilon > + < h\,,\, \tilde{u}^\varepsilon - u > + < h - k\,,\, u >)(t) \end{aligned}$$

which implies that

$$
(6.23)\qquad \begin{cases} \|F^\varepsilon(t) - < k\,,\, u > (t)\|_{C^0(0,T)} \\ \leq \|k^\varepsilon - k\|_{C^0(0,T;H^{-1}(\Omega))}\,\|\tilde{u}^\varepsilon\|_{C^0(0,T;H^1_0(\Omega))} \\ + \|k-h\|_{C^0(0,T;H^{-1}(\Omega))}\,(\|\tilde{u}^\varepsilon\|_{L^\infty(0,T;H^1_0(\Omega))} + \|u\|_{L^\infty(0,T;H^1_0(\Omega))}) \\ + \|h\|_{C^0(0,T;L^2(\Omega))}\,\|\tilde{u}^\varepsilon - u\|_{C^0(0,T;L^2(\Omega))}. \end{cases}
$$

Fix now h in $C^0(0,T;L^2(\Omega))$ such that $\|k-h\|_{C^0(0,T;H^{-1}(\Omega))}$ is small. In view of the convergence (6.22) of k^ε, of the $L^\infty(0,T;H^1_0(\Omega))$ bound of $\tilde{u}^\varepsilon$ and of its strong convergence in $C^0(0,T;L^2(\Omega))$ (recall that $\partial\tilde{u}^\varepsilon/\partial t$ is bounded in $L^\infty(0,T;L^2(\Omega))$), $F^\varepsilon - < k\,,\, u >$ converges strongly to 0 in $C^0(0,T)$. This completes the proof of (6.12).

Theorem 5.1 is proved. □

PROOF OF THEOREM 5.2. Decomposition (5.17) and the convergence result (5.23) on $(\tilde{u}^\varepsilon\,,\,\tilde{\tau}^\varepsilon)$ immediately implies (5.27). This result can also be obtained through direct application of Theorem 3.1 since

$$
(6.24)\qquad \begin{cases} v^\varepsilon(0) = a^\varepsilon - \tilde{a}^\varepsilon \rightharpoonup 0 \text{ weakly in } H^1_0(\Omega) \\ \rho^\varepsilon\,\dfrac{\partial v^\varepsilon}{\partial t}(0) = \rho^\varepsilon(b^\varepsilon - b^0) \rightharpoonup \bar{\rho}(b^0 - b^0) = 0 \text{ weakly in } L^2(\Omega) \\ \sigma^\varepsilon(0) = d^\varepsilon - \tilde{d}^\varepsilon \rightharpoonup d^0 - d^0 = 0 \text{ weakly in } L^2(\Omega) \end{cases}
$$

(see (3.17), (3.18), (3.19), (5.19), (5.21)).

The conservation of the energy (5.16b), namely

$$
\eta^\varepsilon(t) = H^\varepsilon \text{ in } (0,T)
$$

together with the definitions (5.15) of η^ε and H^ε, clearly implies that (5.28) is equivalent to (5.32), which in turn is equivalent to the strong convergence in (5.27).

Theorem 5.2 is proved. □

PROOF OF REMARK 5.3. We firstly prove two lower semi-continuity results.

Since

$$
(6.25)\qquad \begin{cases} \displaystyle\int_\Omega \rho^\varepsilon\,|b^\varepsilon|^2\,dx = \\ \displaystyle = \int_\Omega \rho^\varepsilon\,|b^0|^2\,dx + 2\int_\Omega \rho^\varepsilon\,b^0(b^\varepsilon - b^0) + \int_\Omega \rho^\varepsilon|b^\varepsilon - b^0|^2\,dx \end{cases}
$$

we deduce from the definition (3.18) of b^0 that

$$
(6.26)\qquad \liminf_{\varepsilon\to 0}\int_\Omega \rho^\varepsilon\,|b^\varepsilon|^2\,dx \geq \int_\Omega \bar{\rho}\,|b^0|^2\,dx
$$

where equality takes place if and only if

$$
(b^\varepsilon - b^0) \rightarrow 0 \text{ strongly in} L^2(\Omega).
$$

Further

(6.27)

$$\left\{\begin{aligned}&\int_\Omega [A^\varepsilon \, grad\, a^\varepsilon \, grad\, a^\varepsilon + \beta^\varepsilon |c^\varepsilon|^2] dx\\ &= \int_\Omega [A^\varepsilon \, grad(\tilde{a}^\varepsilon + (a^\varepsilon - \tilde{a}^\varepsilon)) grad(\tilde{a}^\varepsilon + (a^\varepsilon - \tilde{a}^\varepsilon)) + \beta^\varepsilon |c^0 + (c^\varepsilon - c^0)|^2] dx\\ &= \int_\Omega [A^\varepsilon \, grad\, \tilde{a}^\varepsilon \, grad\, \tilde{a}^\varepsilon + \beta^\varepsilon |c^0|^2] dx\\ &+ 2\int_\Omega [A^\varepsilon \, grad\, \tilde{a}^\varepsilon \, grad\,(a^\varepsilon - \tilde{a}^\varepsilon) + \beta^\varepsilon \, c^0 (c^\varepsilon - c^0)] dx\\ &+ \int_\Omega [A^\varepsilon \, grad(a^\varepsilon - \tilde{a}^\varepsilon) grad(a^\varepsilon - \tilde{a}^\varepsilon) + \beta^\varepsilon |c^\varepsilon - c^0|^2] dx\\ &\equiv I^\varepsilon + 2\, II^\varepsilon + III^\varepsilon.\end{aligned}\right.$$

But in view of (6.5) we have

$$I^\varepsilon \to \int_\Omega [A^0 \, grad\, a^0 \, grad\, a^0 + (\bar{\beta} + \kappa)|c^0|^2] dx \tag{6.28a}$$

while clearly

$$\limsup_{\varepsilon \to 0} III^\varepsilon \geq 0. \tag{6.28b}$$

On the other hand

$$\begin{aligned} II^\varepsilon = &\int_\Omega (A^\varepsilon \, grad\, \tilde{a}^\varepsilon - c^0 \, \gamma^\varepsilon) \, grad(a^\varepsilon - \tilde{a}^\varepsilon) dx\\ &+ \int_\Omega c^0 (\beta^\varepsilon \, c^\varepsilon + \gamma^\varepsilon \, grad\, a^\varepsilon) dx\\ &- \int_\Omega c^0 (\beta^\varepsilon \, c^0 + \gamma^\varepsilon \, grad\, \tilde{a}^\varepsilon) dx. \end{aligned}$$

Application of the div-curl lemma implies that the first term tends to zero, because $A^\varepsilon \, grad\, \tilde{a}^\varepsilon - c^0 \, \gamma^\varepsilon$ has a fixed divergence in H^{-1} by (3.12) and (5.11) and $grad(a^\varepsilon - \tilde{a}^\varepsilon)$ tends to 0 weakly in $(L^2(\Omega))^N$ by (6.1). From (3.19) and (6.3) we conclude that

$$II^\varepsilon \to 0. \tag{6.28c}$$

This yields the second lower semicontinuity result, namely

$$\left\{\begin{aligned} &\liminf_{\varepsilon \to 0} \int_\Omega [A^\varepsilon \, grad\, a^\varepsilon \, grad\, a^\varepsilon + \beta^\varepsilon |c^\varepsilon|^2] dx\\ &\geq \int_\Omega [A^0 \, grad\, a^0 \, grad\, a^0 + (\bar{\beta} + \kappa)|c^0|^2 \, dx \end{aligned}\right. \tag{6.29}$$

where equality takes places if and only if

$$\begin{cases} a^\varepsilon - \tilde{a}^\varepsilon \to 0 \text{ strongly in } H_0^1(\Omega) \\ c^\varepsilon - c^0 \to 0 \text{ strongly in } L^2(\Omega). \end{cases}$$

In view of (6.26), (6.29), (5.30) is proved as well as the equivalence between (5.28) and (5.29). The equivalence between (5.32) and (5.28) has already been proved in the proof of Theorem 5.2. Finally (5.31) is equivalent to (5.29) by using in (5.2) an argument similar to (6.6), (6.7).

This completes the proof of Remark 5.3. □

REFERENCES.

[1] A. Bensoussan, J.-L. Lions, G. Papanicolaou. *Asymptotic Analysis for Periodic Structures*, North-Holland, Amsterdam, 1978.

[2] S. Brahim-Otsmane. *Homogénéisation en thermoélasticité*, Thèse de 3ème cycle, Université Pierre et Marie Curie, Paris, 1988.

[3] S. Brahim-Otsmane, G.A. Francfort, F. Murat. *Correctors for the homogenization of the wave and heat equations*, to appear.

[4] G.A. Francfort. *Two variational problems in thermoelasticity*, Ph.D. Thesis, Stanford University, Stanford, 1982.

[5] G.A. Francfort. *Homogenization and fast oscillations in linear thermoelasticity*, in *Numerical Methods for Transient and Coupled Problems*, R.W. Lewis, E. Hinton, P. Bettess, B.A. Schrefler eds., Pineridge Press, Swansea, 1984, pp. 382-393.

[6] G.A. Francfort. *Homogenization and linear thermoelasticity*, SIAM J. Math. Anal., 14 (1984), pp. 696-708.

[7] G.A. Francfort, P. Suquet. *Homogenization and mechanical dissipation in thermoviscoelasticity*, Arch. Rat. Mech. Anal., 96 (1986), pp. 265-293.

[8] J.-L. Lions, E. Magenes. *Problèmes aux Limites Non Homogènes et Applications Vol. 1*, Dunod, Paris, 1968.

[9] J. Marsden, T.J.R. Hughes. *Topics in the mathematical foundations of elasticity*, in *Nonlinear Analysis and Mechanics, Heriot-Watt Symposium Vol. 2*, R.J. Knops ed., Research Notes in Mathematics 27, Pitman, London, 1978, pp. 30-285.

[10] M.L. Mascarenhas. *Homogenisation of a viscoelastic equation with non-periodic coefficients*, Proc. Royal Soc. Edinburgh, 106A (1987), pp. 143-160.

[11] F. Murat. *H-convergence*, in *Séminaire d'Analyse Fonctionnelle et Numérique*, 1977-1978, Université d'Alger, multigraphed.

[12] F. Murat. *Compacité par compensation*. Ann. Sc. Norma. Sup. Pisa, 5 (1978), pp. 489-507.

[13] J.L. Nowinski. *Theory of Thermoelasticity*, Noordhoff International, Alphen aan den Rijn, 1978.

[14] E. Sanchez-Palencia. *Non Homogeneous Media and Vibration Theory*, Lecture Notes in Physics 127, Springer-Verlag, Berlin, 1980.

[15] J. Simon. *Compact sets in $L_p(0,T;B)$*, Ann. Mat. Pura Appl., 146 (1987), pp. 65-96.

[16] S. Spagnolo. *Sulla convergenza di soluzioni di equazioni paraboliche ed ellitiche*, Ann. Sc. Norm. Sup. Pisa, 22 (1968), pp. 577-597.

[17] L. Tartar. *Cours Peccot*, Collège de France, Paris, 1977.

[18] L. Tartar. *Compensated compactness and applications to partial differential equations*, in *Nonlinear Mechanics and Analysis, Heriot-Watt Symposium Vol. 4*, R.J. Knops ed., Research Notes in Mathematics 39, Pitman, London, 1979, pp. 136-212.

[19] L. Tartar. *Nonlocal effects induced by homogenization*, in *Partial Differential Equations and the Calculus of Variations : Essays in Honor of Ennio De Giorgi*, Birkhaüser, Boston, to appear.

CHAPTER 3

Critical Behaviour of Effective Transport Properties Determined from Nearest Neighbour Interactions

L. Poladian*

Abstract. Two and three dimensional composites having the cermet topology are considered in the regime where the volume fraction of the inclusions is close to the critical packing fraction and the contrast of the transport coefficients (dielectric constants in the analysis here) of the phases is large. The critical behaviour of composites having non-intersecting cylindrical (two dimensions) or spherical (three dimensions) inclusions is obtained by assuming that it is determined solely by the arrangement and distribution of nearest neighbour inclusions.

The interaction of pairs of dielectric cylinders or spheres is analyzed using a variation upon the classical method of images. Functional equations are derived and solved to find asymptotic expressions for the induced electrostatic moments. Applying these results to a given nearest neighbour configuration yields an asymptotic approximation to the effective dielectric constant. Using this as an initial approximation, more accurate numerical results can be obtained by using field matching techniques.

The method is applied in detail to several composites having a periodic structure. The corresponding critical behaviour for certain classes of random structures is also presented.

1. Introduction. A useful model of composites having the cermet topology is of a uniform background medium containing non-intersecting cylindrical or spherical inclusions with dielectric constants different from that of the background. Before considering structures with many spheres or cylinders we will first obtain an analytic solution to the problem of two spheres or cylinders in an *arbitrary* external field. The method used is a generalization of the classical technique of images, whereby the field produced by a charge near a circular boundary can be represented as being

* Sir Frank Packer Theoretical Physics Department, University of Sydney, NSW, 2006, Australia.
This work was supported by a Commomwealth Postgraduate Research Award.

produced by two charges: the original charge and an image charge placed on the other side of the circular boundary with a magnitude and position determined by simple rules.

First consider an arbitrary distribution of charge $\sigma(\boldsymbol{r})$ outside a cylinder of radius a and dielectric constant ϵ, the potential produced by this charge alone is denoted by $V_0(\boldsymbol{r}) = V_0(r,\theta)$. Observe now that the following function defined in terms of the potential above, not only satisfies Laplace's equation but also satisfies the boundary conditions at the surface.

$$V(r,\theta) = V_0(r,\theta) - \tau V_0(a^2/r,\theta) \quad r > a \tag{1}$$

$$= \frac{2\epsilon_b}{\epsilon + \epsilon_b} V_0(r,\theta) \quad r < a \tag{2}$$

where

$$\tau = \frac{\epsilon - \epsilon_b}{\epsilon + \epsilon_b} \tag{3}$$

and ϵ_b is the background dielectric constant. The additional term in the exterior field can be regarded as being due to an *image* charge distribution given by

$$\sigma_{\rm im}(r,\theta) = -\tau \frac{a^2}{r^2} \sigma(a^2/r,\theta) \tag{4}$$

In three dimensions it has not been found possible to find an expression for the image field in terms of the original potential except in the case where the dielectric contrast ϵ/ϵ_b is infinite. In this case the exterior field is given by

$$V(r,\theta,\phi) = V_0(r,\theta,\phi) - \frac{a}{r} V_0(a^2/r,\theta,\phi) \tag{5}$$

The interior field is of course zero.

However, the image charge can always be expressed in terms of the original charge even for finite contrast

$$\sigma_{\rm im}(r,\theta,\phi) = -\tau \frac{a}{r}\sigma(a^2/r,\theta,\phi) - \frac{\epsilon_b}{\epsilon+\epsilon_b}\tau \int_a^r \frac{a}{\rho}\left(\frac{\rho}{r}\right)^{-\frac{1}{2}(\tau+1)} \sigma(a^2/\rho,\theta,\phi)\, d\rho \tag{6}$$

A derivation of this result can be found in Poladian [1].

2. Pair Interactions. In the case of two spheres or cylinders, each will have its own image charge distribution, and these charge distributions will have to be mutual images in order to *simultaneously* satisfy the boundary conditions on both surfaces. Before presenting the coupled equations for the mutual images a number of simplifications can be made. In the case of cylinders, it is sufficient to consider charge distributions confined to the axis of symmetry (i.e. line charge distributions only). This is also true for spheres if the external field has azimuthal symmetry about the line joining the spheres. However, even in the general case a line charge distribution is sufficient to obtain the asymptotic solution, since only those components of the external field which have azimuthal symmetry contribute asymptotically. Also, the complicated term involving the integral can be neglected for large contrast because it is of smaller order than the other terms. These results are proved in detail elsewhere [1].

Consider a pair of cylinders or spheres with radii a and a' and dielectric constants ϵ and ϵ' respectively. The distance between their centres is d and let h denote the size of the gap. If we let the image charge distributions be $\sigma(x)$ and $\sigma'(x')$ respectively, where x and x' are measured from the centre of the respective cylinder or sphere towards the other, then the following pair of coupled functional equations is obtained.

$$\sigma'(\frac{a'^2}{d-x}) = -\tau'(\frac{d-x}{a'})^{4-D}\sigma(x) \tag{7}$$

$$\sigma(\frac{a^2}{d-x'}) = -\tau(\frac{d-x'}{a})^{4-D}\sigma'(x') \tag{8}$$

where

$$\tau = \frac{\epsilon-\epsilon_b}{\epsilon+\epsilon_b}, \qquad \tau' = \frac{\epsilon'-\epsilon_b}{\epsilon'+\epsilon_b} \tag{9}$$

D denotes the dimensionality: $D = 2$ for cylinders and $D = 3$ for spheres. A solution to these equations for the case $D = 2$ has been given by McPhedran *et al.* [2] and for the case $D = 3$ by Poladian [1]. It was subsequently discovered that the equations possess a complete, orthogonal set of solutions.

The functional equations have fixed points at

$$x = a_\infty \quad \text{and} \quad x' = a'_\infty \tag{10}$$

$$x = d - a'_\infty \quad \text{and} \quad x' = d - a_\infty \tag{11}$$

which are the roots of the coupled quadratic equations

$$x(d-x') = a^2 \tag{12}$$

$$x'(d-x) = a'^2 \tag{13}$$

Solutions to the functional equations can be found by studying the behaviour near these fixed points.

The complete set of solutions to the equations is then given by

$$\sigma_n(x) = \tau^{\frac{1}{2}}(\frac{a_\infty}{a})^{-s_n}(a_\infty - x)^p(d - a'_\infty - x)^q \tag{14}$$

$$\sigma'_n(x') = -(-1)^n\tau'^{\frac{1}{2}}(\frac{a'_\infty}{a'})^{-s_n}(a'_\infty - x')^p(d - a_\infty - x')^q \tag{15}$$

where

$$p = s_n - 2 + D/2 \quad \text{and} \quad q = -s_n - 2 + D/2 \tag{16}$$

and

$$s_n = \frac{1}{2}\frac{\log(\tau\tau') + 2in\pi}{\log(a_\infty a'_\infty/aa')} \tag{17}$$

The general solution is then

$$\sigma(x) = \sum_{n=-\infty}^{\infty} A_n \sigma_n(x) \tag{18}$$

$$\sigma'(x') = \sum_{n=-\infty}^{\infty} A_n \sigma'_n(x') \tag{19}$$

The coefficients A_n are determined by the details of the external field.

If we wish to use the above results in conjunction with other techniques such as field expansions then we can convert the image charges to their corresponding electrostatic multipole moments. The moments are given by

$$Q_l = \int_0^{a_\infty} x^l \sigma(x) dx \tag{20}$$

$$Q'_l = -(-1)^l \int_0^{a'_\infty} x'^l \sigma'(x') dx' \tag{21}$$

In all further calculations we will retain only the asymptotic part: the part which diverges as the gap size goes to zero and the contrast goes to infinity. The superiority of the above expansion now becomes apparent, since out of all the terms in the series expansion of the general solution only the term $n = 0$ contributes to the asymptotic behaviour. This has the important implication that the asymptotic solution to all problems with cylinders or spheres of given sizes and separations is the same, independent of the detail or nature of the external field, up to a scale factor A_0, which can be determined by normalisation. Henceforth, only the $n = 0$ term will be retained in any calculations and s_0 will be abbreviated to s.

The multipole moments can be evaluated in terms of the hypergeometric function $F(a, b; c; z)$ [3]

$$\begin{aligned} Q_l &= \frac{A_0\sqrt{\tau}(a_\infty/a)^{s+1}a_\infty^l}{a^{3-D}} \frac{\Gamma(l+1)\Gamma(s-1+D/2)}{\Gamma(l+s+D/2)} \\ &\times F(l+1, s+2-D/2; l+s+D/2; a_\infty^2/a^2), \end{aligned} \tag{22}$$

$$\begin{aligned} Q'_l &= \frac{A_0\sqrt{\tau'}(a'_\infty/a')^{s+1}a'^l_\infty}{-(-1)^l a'^{3-D}} \frac{\Gamma(l+1)\Gamma(s-1+D/2)}{\Gamma(l+s+D/2)} \\ &\times F(l+1, s+2-D/2; l+s+D/2; a'^2_\infty/a'^2), \end{aligned} \tag{23}$$

The constant A_0 is found by calculating the potential difference between the centres of the nearest neighbours. It can also be expressed in terms of hypergeometric functions but the full expression is very complicated. The asymptotic behaviour (small h and large ϵ) is given by

$$\Delta V = \frac{A_0}{2\pi\epsilon_b s}\sqrt{\frac{2}{aa'}}\left[1 - 2s\{\frac{1}{2}\log(\frac{2h}{d}) + \psi(s+1) + \gamma + \frac{\epsilon-\epsilon'}{\epsilon+\epsilon'}\log(\frac{a}{a'})\}\right] \tag{24}$$

for two dimensions and

$$\Delta V = \frac{A_0}{4\pi\epsilon_b}\frac{d}{aa'} \tag{25}$$

for three dimensions.

Further, in two dimensions, when there is more than one nearest neighbour the potential difference between cylinder A and its neighbour B is modified by the interaction between A and another neighbour C. This change is comparable to the potential difference itself and must be retained to obtain the correct asymptotic behaviour. The corresponding effect in three dimensions is of smaller order and vanishes asymptotically. If θ is the angle between the line joining the centres of A and B and a similar line through A and C the potential difference is increased by an amount

$$\Delta V(\theta) = \frac{A_0}{2\pi\epsilon_b}\sqrt{\frac{2}{aa'}}[2\log(1-e^{i\theta})] \tag{26}$$

An important quantity is the polarizability per sphere or cylinder which is defined as

$$P(a,a',\epsilon,\epsilon') = \frac{Q_1+Q_1'}{2\epsilon_b d^{D-1}\Delta V} \tag{27}$$

For a pair of cylinders we have

$$P(a,a',\epsilon,\epsilon') = \pi\left\{2\sqrt{\frac{2hd}{aa'}} + (\frac{\epsilon_b}{\epsilon}+\frac{\epsilon_b}{\epsilon'})[\log(\frac{d}{2h}) - 2\psi(s+1) - 2\gamma] + \frac{1}{2}(\frac{\epsilon_b}{\epsilon}-\frac{\epsilon_b}{\epsilon'})\log(\frac{a'}{a})\right\}^{-1} \tag{28}$$

and for spheres

$$P(a,a',\epsilon,\epsilon') = 2\pi\frac{aa'}{d^2}\left\{\frac{1}{2}\log(\frac{d}{8h}) - \psi(s+\frac{1}{2})\right\} \tag{29}$$

$$\sim -2\pi\frac{aa'}{d^2}\log\left[(\frac{\epsilon_b}{\epsilon}+\frac{\epsilon_b}{\epsilon'})\sqrt{\frac{4aa'}{d^2}} + \sqrt{\frac{2h}{d}}\right] \tag{30}$$

The above results are the asymptotic solutions in terms of both the gap size and the contrast. If we examine the two limits separately, much simpler results are obtained which can be presented in a form similar to critical exponents. The results for $D = 1$ (laminar structures) are also given. In the following table the gap size has been written in terms of a volume or area fraction f which approaches a critical fraction f_c as the spheres or cylinders touch.

Note the constant γ in the denominator of the two dimensional result. The leading order singularity is $\epsilon/\log\epsilon$, but there are also confluent singularities the first of which is $\epsilon/(\log\epsilon)^2$ which must be included to obtain the full singularity. The correct choice of the constant in the denominator achieves this. Further, the value of this constant depends on the number and orientation of nearest neighbours and can be determined in the general case from equations (24) and (26). The above special

Table 1: Critical Behaviour of the polarizability of a pair of spheres or cylinders in the two limits of infinite dielectric contrast as the gap between them goes to zero $f \to f_c$, or in the touching limit as the contrast goes to infinity

	$\epsilon = \epsilon' = \infty$	$f = f_c$
$D = 1$	$(f_c - f)^{-1}$	ϵ/ϵ_b
$D = 2$	$(f_c - f)^{-\frac{1}{2}}$	$\dfrac{\epsilon/\epsilon_b}{\log(\epsilon/\epsilon_b) - \gamma}$
$D = 3$	$\log(f_c - f)$	$\log(\epsilon/\epsilon_b)$

cases agree with the results obtained by Batchelor & O'Brien [4] using variational arguments. However, no one else has previously given the correct formula for the uniform interpolation between these special cases.

3. General Structures. In a general structure with a large number of spheres or cylinders, the effective dielectric constant can be obtained from the polarizability per unit volume. This polarizability will contain contributions from every nearest neighbour pair. The contribution from the interaction between the ith and jth spheres or cylinders (assuming them to be nearest neighbours) is given by

$$\epsilon(i,j) \sim d_{ij}^{D-1}(V_{ij}\cos\Theta_{ij}/E)P(a_i, a_j, \epsilon_i, \epsilon_j) \tag{31}$$

where d_{ij} is the distance between their centres, V_{ij} is the potential difference between their centres produced by a uniform external field E, and Θ_{ij} is the angle between the line joining the centres and the direction of the particular component of the polarization vector we are interested in.

Then, the component of the effective dielectric tensor which relates the above component of the polarization to the given applied field direction is

$$\frac{\epsilon_{\text{eff}}}{\epsilon_b} \sim \sum_{i=1}^{M} n_i \sum_{j=1}^{N_i} \epsilon(i,j) \tag{32}$$

where N_i is the number of nearest neighbours around the ith inclusion, and n_i is the number of inclusions per unit volume. The first summation must be done over a sufficiently large number M, of inclusions such that a macroscopic average can be obtained. The above formula is extremely general and can be applied to many different structures. Before any further progress can be made with the expression specific information is required about the structure. For a random structure, this means information about the distribution of gap sizes near percolation and the potential differences across these gaps. For a regular structure, it involves a knowledge of the number of basis elements and their relative positions. The validity and limitations of the asymptotic analysis will be tested by investigating a number of regular structures for which there exist alternative methods for rigorously calculating the effective dielectric constants.

4. Periodic Structures. The potential difference between two inclusions is related to the external field and the distance between the centres by the formula

$$V_{ij} = Ed_{ij}\cos\Phi_{ij} \tag{33}$$

where Φ_{ij} is the angle made between the line joining the centres and the direction of the applied field. This is not necessarily the same angle as Θ_{ij} since for an anisotropic structure the induced polarization need not be in the same direction as the applied field.

The effective dielectric constant is then given by

$$\frac{\epsilon_{\text{eff}}}{\epsilon_{\text{b}}} \sim \sum_{i=1}^{M}\sum_{j=1}^{N_i} n_i d_{ij}^D P(a_i, a_j, \epsilon_i, \epsilon_j)\cos\Theta_{ij}\cos\Phi_{ij} \tag{34}$$

where M is now the number of basis elements in a unit cell. The above formula applies to any periodic structure.

Consider structures in which all nearest neighbours are equivalent and isotropically distributed. Such a structure has an isotropic effective dielectric constant and the average value of $\cos\Theta_{ij}\cos\Phi_{ij}$ is one half in two dimensions and one third in three dimensions. The formula is simplified to

$$\frac{\epsilon_{\text{eff}}}{\epsilon_{\text{b}}} \sim \sum_{i=1}^{M} n_i d_{ij}^D \frac{N_i}{D} P(a_i, a_j, \epsilon_i, \epsilon_j) \tag{35}$$

The results for the lattices given in table 2 can all be obtained from the above formula.

Table 2: Coordination and geometric parameters for various lattice structures which can be analysed using the asymptotic formula for the effective dielectric constant.

LATTICE	N_i	n_i	d_{ii}
Square	4	1	1
Hexagonal	6	$2/\sqrt{3}$	1
SC	6	1	1
BCC	8	2	$\sqrt{3}/2$
FCC	12	4	$1/\sqrt{2}$

5. Field Expansion Methods. In order to examine the accuracy and limitations of the above asymptotic results, a rigorous method is used to calculate the induced multipole moments in a periodic configuration of spheres or cylinders.

The charge distribution on the surface of each cylinder or sphere is expanded in an appropriate set of orthogonal functions. In two dimensions we have

$$\sigma^i(\theta) = \sum_{l=-\infty}^{\infty}{}' Q_l^i \frac{\exp(il\theta)}{|l|} \tag{36}$$

where the $l = 0$ term does not appear since each cylinder remains charge neutral. In three dimensions

$$\sigma^i(\theta, \phi) = \sum_{l=0}^{\infty} \sum_{m=-l}^{l} Q_{lm}^i C_{lm}(\theta, \phi) \tag{37}$$

where $C_{lm}(\theta, \phi)$ is the unnormalized surface spherical harmonic [5].

The fields throughout space are calculated from the charge distributions and any external fields, and then required to satisfy the boundary conditions at the surfaces of the cylinders or spheres. By applying the orthogonality conditions and addition theorems [6] for the harmonic functions a matrix equation can be obtained for the unknown coefficients, Q. In two dimensions the equation is

$$\sum_{k=-\infty}^{\infty} \sum_j M_{l,i}^{k,j} Q_k^j = 2\pi\epsilon_b V_l^i, \tag{38}$$

where

$$M_{l,i}^{k,j} = (-1)^l \binom{|l|+|k|}{|l|} Y_{l+k}(\boldsymbol{r}_i - \boldsymbol{r}_j) + \frac{\delta_{kl}\delta_{ij}(\epsilon_i + \epsilon_b)}{a_i^{2l}(\epsilon_i - \epsilon_b)} \tag{39}$$

and

$$Y_l(\boldsymbol{r}) = r^{-|l|} \frac{\exp(il\theta)}{|l|}, \qquad \boldsymbol{r} = (r, \theta). \tag{40}$$

The external field is given by

$$V_{ext}(\boldsymbol{r}) = \sum_l V_l^i r^{|l|} \frac{\exp(il\theta)}{|l|} \tag{41}$$

In three dimensions the result is

$$\sum_{k=0}^{\infty} \sum_{n=-k}^{k} \sum_j M_{lm,i}^{kn,j} Q_{kn}^j = 4\pi\epsilon_b V_{lm}^i, \tag{42}$$

where

$$\begin{aligned} M_{lm,i}^{kn,j} &= (-1)^{l+m} \sqrt{\binom{l+k-n+m}{l+m}\binom{l+k+n-m}{k+n}}\, Y_{l+k,m-n}(\boldsymbol{r}_i - \boldsymbol{r}_j) \\ &+ \frac{\delta_{kl}\delta_{mn}\delta_{ij}(l\epsilon_i + l\epsilon_b + \epsilon_b)}{a_i^{2l+1}(l\epsilon_i - l\epsilon_b)} \end{aligned} \tag{43}$$

and

$$Y_{lm}(\boldsymbol{r}) = r^{-l-1} C_{lm}(\theta, \phi), \qquad \boldsymbol{r} = (r, \theta, \phi). \tag{44}$$

and the external field is

$$V_{ext}(\boldsymbol{r}) = \sum_{lm} V_{lm}^{i} r^{l} C_{lm}(\theta, \phi) \tag{45}$$

The full details of the derivation of these matrix equations have been given by Perrins *et al.* [8] for two dimensions and Poladian & McPhedran [7] for three dimensions.

If the matrix equation is solved to obtain the unknown coefficients Q_l^i or Q_{lm}^i in the case of a uniform external field $\boldsymbol{E}$ the effective dielectric tensors can be calculated. In two dimensions

$$\begin{pmatrix} \frac{1}{2V}\sum_i (Q_1^i + Q_{-1}^i) \\ \frac{1}{2iV}\sum_i (Q_1^i - Q_{-1}^i) \end{pmatrix} = (\epsilon^{\text{eff}} - \epsilon_{\text{b}} I) \begin{pmatrix} E_x \\ E_y \end{pmatrix} \tag{46}$$

and in three dimensions

$$\begin{pmatrix} \frac{1}{\sqrt{2}V}\sum_i (Q_{1,1}^i + Q_{1,-1}^i) \\ \frac{1}{\sqrt{-2}V}\sum_i (Q_{1,1}^i - Q_{1,-1}^i) \\ \frac{1}{V}\sum_i Q_{1,0}^i \end{pmatrix} = (\epsilon^{\text{eff}} - \epsilon_{\text{b}} I) \begin{pmatrix} E_x \\ E_y \\ E_z \end{pmatrix} \tag{47}$$

Although this method is rigorous, the size of the matrix required to obtain accurate answers increases very rapidly [7] as the volume fraction and contrast increase. The asymptotic solution allows us to reduce the size of the matrix by first extracting the knowledge we have about the behaviour of the high order multipoles. Let us write

$$Q_l = Q_l^{\text{n.n.}} + Q_l^{\text{res.}} \tag{48}$$

where the first term is obtained from the asymptotic analysis and represents the nearest neighbour interactions. The second term represents the residual interactions and should be very small for very high order multipoles (l large), thus any corrections to the asymptotic results can be obtained by inverting a relatively small matrix as compared to the matrix inversion required when the asymptotic behaviour is not known.

6. Two Dimensional Structures. The simplest two dimensional structure is the square array of cylinders. The effective dielectric constant $\frac{\epsilon_{\text{eff}}}{\epsilon_{\text{b}}}$ is a function of two variables a_i and ϵ_i. The asymptotic analysis predicts

$$\frac{\epsilon_{\text{eff}}}{\epsilon_{\text{b}}} = \frac{\pi}{\sqrt{\frac{2h}{d} + \frac{\epsilon_{\text{b}}}{\epsilon_i}[\log(d/2h) - 2\gamma - 2\psi(s+1)]}} + O(1) \tag{49}$$

with $h = d - 2a_i$.

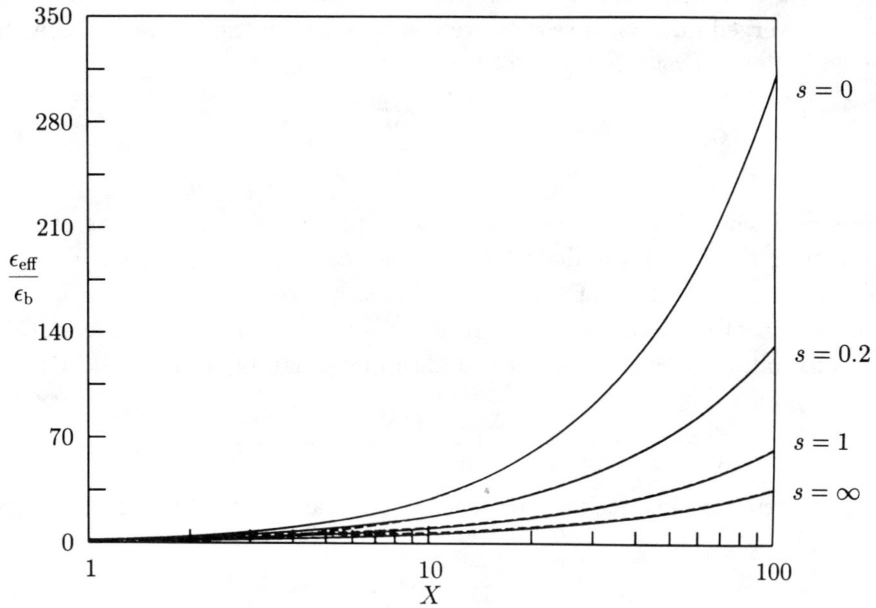

Figure 1: Effective dielectric constant for the square lattice as a function of the asymptotic variable X for various values of s. The solid line represents the exact result obtained by numerical inversion of a sufficiently large matrix. The dashed line represents the asymptotic result combined with a numerical estimate of the additive constant.

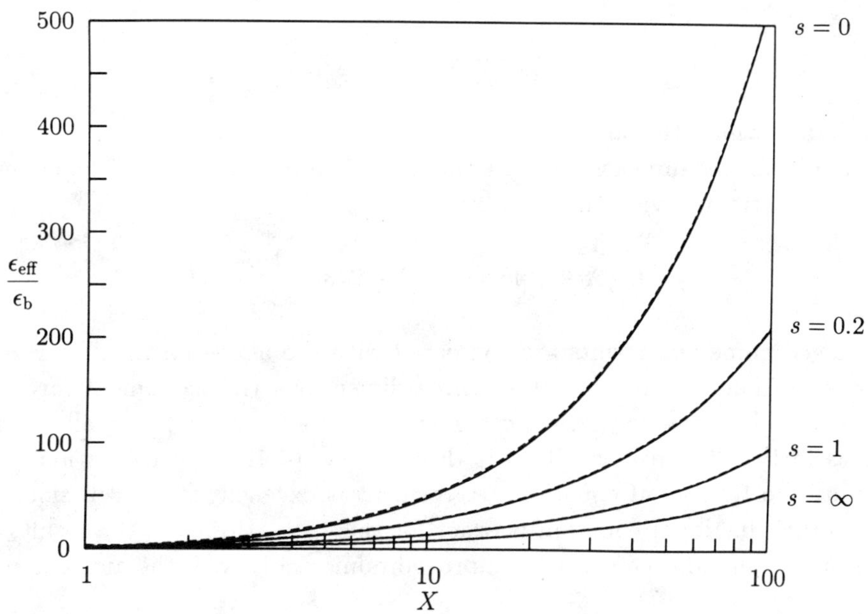

Figure 2: Effective dielectric constant for the hexagonal structure. The functional form is the same as for the square lattice apart from a different multiplicative factor and a change in the constant term in the denominator.

The matrix equation is solved numerically with the aid of the asymptotic solution. The converged numerical results are compared with the above formula. Both are plotted in figure 1 as a function of

$$X = \left(\frac{\epsilon_b}{\epsilon_i} + \sqrt{\frac{2h}{d}}\right)^{-1} \tag{50}$$

which has been chosen as a natural variable which tends to infinity as either the gap size h tend to zero or the dielectric contrast tends to infinity. The results are compared for various values of s which is a measure of which factor (gap size or dielectric contrast) is more important in determining asymptotic behaviour. The additive constant of order unity is also found (approximately) and the above formula becomes

$$\frac{\epsilon_{\text{eff}}}{\epsilon_b} \sim \frac{\pi(s+1)X}{1 + 2s[\log X + \log(s+1) - \psi(s) - \gamma]} - 3 \tag{51}$$

The next structure investigated is the hexagonal array of cylinders. The asymptotic behaviour is given by

$$\frac{\epsilon_{\text{eff}}}{\epsilon_b} \sim \frac{\sqrt{3}\pi(s+1)X}{1 + 2s[\log X + \log(s+1) - \psi(s+1) - \gamma + \frac{1}{2}\log 3]} - 4 \tag{52}$$

The additive constant of order unity was found numerically as for the square array. No relationship is known between the two additive constants for the square and hexagonal arrays. The results for the hexagonal structure are shown in figure 2.

7. Three Dimensional Structures. In three dimensions we chose

$$X = \left(\frac{2\epsilon_b}{\epsilon'} + \sqrt{\frac{2h}{d}}\right)^{-1} \tag{53}$$

as the natural asymptotic variable.

The simplest three dimensional structure is the simple cubic array of spheres. The asymptotic behaviour is given by

$$\frac{\epsilon_{\text{eff}}}{\epsilon_b} \sim \pi \log X + \pi[\log(s + \frac{1}{2}) - \psi(s + \frac{1}{2})] - 3.6 \tag{54}$$

The converged numerical results are compared with the above formula for various values of s in figure 3. Note that in three dimensions the agreement for small X worsens as s becomes larger. In other words the agreement for small ϵ is not as good as in two dimensions. This is directly due to the approximation made in obtaining the functional equation. Agreement is excellent for $s = 0$ since the solution to the functional equation is exact for this case. However, the results *are* valid asymptotically and can still be improved numerically with the aid of a *small* matrix.

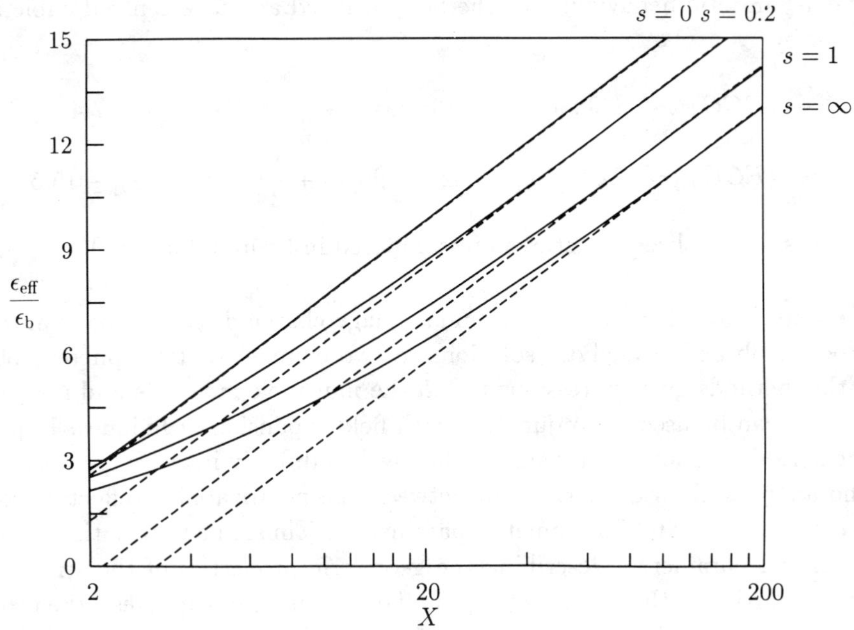

Figure 3: Asymptotic behaviour for the sc lattice. Note that the agreement for small X when s is large is not as good as in two dimensions. However, there is always agreement for large X.

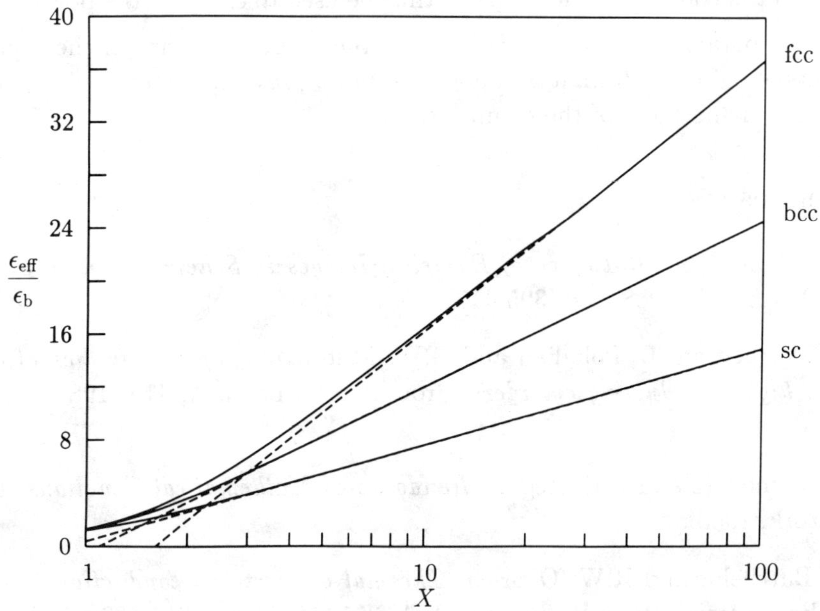

Figure 4: Comparison of different three dimensional structures. All the structures have the same asymptotic form but with different multiplicative factors or slopes.

The asymptotic behaviours for the body centred and face centred cubic arrays are given by

$$\frac{\epsilon_{\text{eff}}}{\epsilon_b}(BCC) \sim \sqrt{3}\pi \log X + \sqrt{3}\pi[\log(s+\frac{1}{2}) - \psi(s+\frac{1}{2})] - 7.4 \qquad (55)$$

$$\frac{\epsilon_{\text{eff}}}{\epsilon_b}(FCC) \sim 2\sqrt{2}\pi \log X + 2\sqrt{2}\pi[\log(s+\frac{1}{2}) - \psi(s+\frac{1}{2})] - 15.5 \qquad (56)$$

The results for all three structures are compared in figure 4 for $s = 0$.

8. Discussion. The use of continuous image charge distributions is an important tool in obtaining analytic solutions to two cylinder or two sphere problems. Since the methods give expressions for *all* the multipole moments and not just the first one, it can be used in conjunction with field expansion techniques. In particular, for periodic structures it removes the need to directly invert large matrices.

The analysis gives a relationship between the polarizability (effective conductance etc.) of a gap and its geometric parameters. Thus, in a percolating structure, if the critical parameters describing the geometric properties of the structure are known (in particular the density of gaps and distribution of gap sizes), then combining this relationship with the asymptotic formulas derived in this paper should give the critical behaviour of the effective dielectric constant (or effective conductivity). In other words, the asymptotic formulas link the geometry to the physical properties and may even provide a scaling law between the geometric critical exponents and those for transport coefficients (conductivity etc.).

Further, the functional equation analysis in two dimensions is exact for all values of ϵ, even in the complex plane, and can thus be used to examine the poles of $\frac{\epsilon_{\text{eff}}}{\epsilon_b}$ in terms of the parameter s and the residues of these poles in terms of the expansion coefficients A_n. Such information about the poles provides a means of calculating the absorption spectrum of the composite.

References

1. L. Poladian, *General Theory of Electrical Images in Sphere Pairs*, Q. J. Mech. Appl. Math., 41 (1988), pp. 395-417.

2. R.C. McPhedran, L. Poladian and G.W. Milton, *Asymptotic studies of closely spaced, highly conducting cylinders*, Proc. R. Soc. Lond. A, 415 (1988), pp. 185-196.

3. M. Abramowitz and I.A. Stegun, *Handbook of Mathematical Functions* Dover, New York, 1965 .

4. G.K. Batchelor and R.W. O'Brien, *Thermal or electrical conduction through a granular material*, Proc. R. Soc. Lond. A, 355 (1977), pp. 313-333.

5. D.M. Brink and G.R. Satchler, *Angular Momentum* 2nd ed. Clarendon Press, Oxford, 1968 p145ff.

6. R.J.A. Tough and A.J. Stone, *Properties of the regular and irregular solid harmonics*, J. Phys. A, 10 (1977), pp. 1261-1269.

7. L. Poladian and R.C. McPhedran, *Effective Transport Properties of Periodic Composite Materials*, Proc. R. Soc. Lond. A, 408 (1986), pp. 45-59.

8. W.T. Perrins, D.R. McKenzie and R.C. McPhedran, *Transport properties of regular arrays of cylinders*, Proc. R. Soc. Lond. A, 369 (1979), pp. 207-225.

CHAPTER 4

Bounds on Various Electrostatic and Hydrodynamic Capacities

Jacob Rubinstein*
Sal Torquato†

Abstract. We are concerned here with estimating the permeability of porous media in terms of rigorous bounds. The problem lay dormant for many years following the pioneering paper of Prager [9]. It was revived in the early eighties by Berryman and Torquato. We first characterize the permeability in terms of a microscopic canonical problem. This is used to formulate variational principles and then bounds. We discuss the relations between the permeability and the inverse effective drag coefficient. An analogy is drawn between the permeability problem and the problem of diffusion-controlled reactions. Finally, we compare our results to estimates derived by effective field approximations.

1. Bounds on the Permeability. Consider a random partition of R^3 into a void region $Y(\omega)$ and a solid region $Y^C(\omega)$. Here ω is a random variable, and the statistics of $Y(\omega)$ is stationary. We assume that $Y(\omega)$ is filled with fluid which satisfies Stokes equations

$$\mu\Delta u = \nabla P , \qquad \nabla\cdot u = 0 \quad \text{in } Y(\omega), \tag{1.1}$$

$$u = 0 \quad \text{on } \partial Y(\omega). \tag{1.2}$$

Here $u(x,\omega)$ is the velocity, μ is the viscosity, and $P(x,\omega)$ is the pressure. Hydrologists replace the random system (1.1)-(1.2) by the deterministic system (Darcy's law):

$$U = -\frac{K}{\mu}\nabla P_0 , \quad \nabla\cdot U = 0 , \tag{1.3}$$

*Department of Mathematics, Technion – Israel Institute of Technology, 32000 Haifa, Israel.

†Department of Mechanical and Aerospace Engineering, North Carolina State University, Raleigh, NC 27695.

where U(x) is the average velocity, ∇P_0 is the applied pressure gradient, and the tensor K is called the ***permeability*** of the medium. While (1.3) was originally written down as an experimental law, it is possible to derive it from (1.1)-(1.2) using a two-scale method. The two scales are a fast scale associated with $\partial Y(\omega)$, and a slow scale associated (for example) with the variations of the force driving the fluid. The two-scale analysis identifies the following canonical problem:

$$\left.\begin{aligned} \Delta W &= \nabla\Pi - E \\ \nabla\cdot W &= 0 \end{aligned}\right\} \text{ in } Y^o(w) \tag{1.4}$$

$$W = 0 \qquad \text{on } \partial Y^o(\omega) \ . \tag{1.5}$$

Here E is the unit dyadic, and $Y^o(\omega)$ is the scaled (with respect to the fast scale) form of $Y(\omega)$. Then, K can be written as

$$K = \langle W\rangle, \tag{1.6}$$

where angular brackets denote an ensemble average (W is defined to be identically zero in Y^C).

Notice that unlike other transport problems in disordered media (conductivity, elasticity, etc.), the effective (homogenized) equation (1.3) is of a different nature than the microscopic equation (1.1). Another way to describe this degeneracy is to say that the permeability K has no microscopic counterpart.

We now formulate upper and lower bounds on K. For simplicity we assume from now on that $Y^o(\omega)$ is isotropic, so the permeability is characterized by a single constant.

Upper Bound. It has been shown by Rubinstein & Torquato [11] that the permeability is bounded from above by

$$K \leqslant \langle \nabla u : \nabla u \ I \rangle, \quad \forall\ u \in A \tag{1.7}$$

where $I(\omega)$ is the characteristic function of $Y(\omega)$, and

$$A = \{\text{stationary smooth } (H^1_{loc}) \text{ functions } u(x,\omega), \text{ s.t.} \tag{1.8}$$
$$\nabla \times (\Delta u + \hat{e}) = 0 \text{ in } Y(\omega), \ \hat{e} \text{ being a unit vector}\}.$$

Lower Bound. It has been shown by Rubinstein & Torquato [11] that k is bounded from below by

$$K \geqslant \left(\langle \nabla u : \nabla u \ I\rangle\right)^{-1}, \quad \forall\ u \in B, \tag{1.9}$$

where

$$B = \{\text{stationary smooth functions } u(x,\omega), \text{ s.t. } u = 0 \tag{1.10}$$
$$\text{on } \partial Y(\omega),\ \nabla\cdot u = 0 \text{ in } Y(\omega), \text{ and } \langle u\cdot\hat{e}\ I\rangle = 1\}.$$

Our task is to construct admissible trial fields, and to perform the necessary averaging. In what follows we describe four different types of admissible fields and their bounds: void, interfacial-surface, multiple-scattering, and security-spheres bounds. We first define the following correlation functions:

$$F_{vv}(r) = \langle I(x)I(x+r)\rangle , \qquad (1.11)$$

$$F_{sv}(r) = \langle M(x)I(x+r)\rangle , \qquad (1.12)$$

$$F_{ss}(r) = \langle M(x)M(x+r)\rangle , \qquad (1.13)$$

where $M(x) = |\nabla I(x)|$. The volume fraction of the void region, ϕ_1, satisfies $\phi_1 = \langle I(x)\rangle$.

A. Void Upper Bound. Let

$$u_A(x,\omega) = \frac{1}{\phi_2} \int \psi(x-y)\cdot\hat{e}\left[I(y)-\phi_1\right]dy. \qquad (1.14)$$

Here $\phi_2 = 1-\phi_1$, and ψ stands for the fundamental solution of Stokes' equation [7]. The trial field $u_A \in A$, together with (2.7), gives us

$$K \leqslant \langle\nabla u_A : \nabla u_A I\rangle \leqslant \langle\nabla u_A : \nabla u_A\rangle = \frac{2}{3\phi_2^2} \int_0^\infty \left(F_{vv}(r)-\phi_1^2\right) r dr. \qquad (1.15)$$

This upper bound was first derived by Berryman and Milton [2] using a different approach.

B. Interfacial-surface Upper Bound. Let

$$u_B(x,\omega) = \int \psi(x-y)\left[\hat{e}I(y)-\xi(y)M(y)\right]dy. \qquad (1.16)$$

Next we define the expected area of ∂Y per unit volume as

$$s = \langle M(x)\rangle, \qquad (1.17)$$

and set

$$\xi = \frac{\phi_1}{s}\hat{e} . \qquad (1.18)$$

Using (2.7) and the inequality $I \leqslant 1$, we find

$$K \leqslant \frac{2}{3}\int_0^\infty \left[\frac{\phi_1^2}{s^2} F_{ss}(r) - \frac{2\phi_1}{s} F_{sv}(r) + F_{vv}(r)\right] r dr. \qquad (1.19)$$

This bound was first obtained by Doi [6] using a different methodology.

C. Multiple Scattering Upper Bound. The first two bounds were rather general. They apply to any microscopic topology. A common theoretical model for porous media is a distribution of spherical obstacles. In this case we can use multiple scattering techniques to establish approximations and even bounds. For this purpose, let $\chi(r)$ be the characteristic function of a sphere of radius a, and let $S(x,y)$ be the single-sphere scattering function (for

Stokes' equation), i.e.,

$$S(x,y) = \left(1 + \frac{a^2}{6}\Delta\right)\psi(x,y). \tag{1.20}$$

Then we define (see also [13])

$$u_C(x,\omega) = \int S(x-y)\cdot\hat{e}dy - \frac{1}{n}\sum S(x-r_i)\cdot\hat{e}\chi\left(|x-r_i|\right), \tag{1.21}$$

where n is the number density of the spheres. In computing the averge in (1.7) we make use of the functions $G_{q+1}(r^q)$ which is the probability density function associated with finding the origin in the void, and a configuration of particle centers $(r_1,\ldots,r_q)$ [12]. Substituting (1.21) in (1.7), we find

$$K \leqslant \frac{1}{2n^2}\left[\int G_2(y)\tau(y):\tau(y)dy + \iint Q_3(y_1,y_2)\tau(y_1):\tau(y_2)dy_1dy_2\right], \tag{1.22}$$

where

$$Q_3(y_1,y_2) = G_3(y_1,y_2) - n\left(G_2(y_1)+G_2(y_2)\right) + n^2\phi_1,$$

and

$$\tau = \nabla(S\cdot\hat{e}) + \nabla(S\cdot\hat{e})^T.$$

D. Security Spheres Lower Bound. Again we treat a random distribution of identical spheres of radius a. To construct a function in class B, we consider for each sphere a security sphere of radius b. If we choose b to be half the distance to its nearest neighbor, then the security spheres do not overlap. In each sphere we construct the appropriate solution to Stokes' equation. Using (1.9) we find [11]

$$K \geqslant \left(6\pi an\int_a^{\infty} f\left(\frac{a}{b}\right)H(b)db\right)^{-1}, \tag{1.23}$$

where H(b) is the nearest neighbor distribution function, and f(λ) is given by

$$f(\lambda) = (1-\lambda^5)\left(1 - \frac{9\lambda}{4} + \frac{5\lambda^3}{2} - \frac{9\lambda^5}{4} + \lambda^6\right)^{-1}. \tag{1.24}$$

Bounds Evaluation. A detailed evaluation of the upper bounds is given in [11]. We give here a few examples. We denote the right hand side of (1.15), (1.19) and (1.22) by K_A, K_B and K_C respectively. For an uncorrelated distribution of identical spheres we get the following dilute expansion:

$$K_A \sim \frac{6}{5}K_s, \tag{1.25}$$

$$K_B \sim K_s\left(1 - \frac{15}{8}\phi_2\right), \tag{1.26}$$

$$K_C \sim K_s\left(1 - \frac{3}{2}\phi_2\right), \tag{1.27}$$

where $K_s = \frac{2a^2}{9\phi_2}$ is the Stokes permeability. We see that the void upper bound does not even capture the dilute Stokes limit. For a distribution of impenetrable spheres we obtain:

$$K_A \sim \frac{6}{5} K_S , \tag{1.28}$$

$$K_B \sim K_S(1 - 5\phi_2), \tag{1.29}$$

$$K_C \sim K_S(1 - 5.7\phi_2), \tag{1.30}$$

We also present (Figure 1) a numerical computation of K_B, an improved version of K_C, and the Kozeny-Carman empirical formula [1].

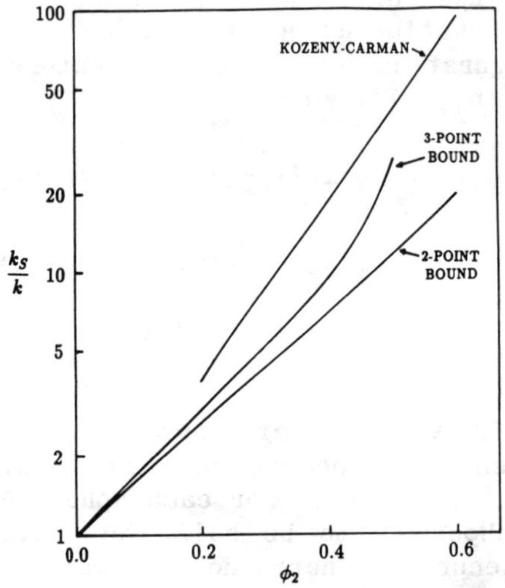

Figure 1: Comparison of the evaluation of the two-point bound (1.19) and the three-point bound (1.22) for impenetrable spheres.

The evaluation of bound D (the lower bound) involves an intrinsic difficulty. The function $f(\frac{a}{b})$ has a pole at $a = b$. Therefore the bound is nontrivial only if $H(b)$ vanishes rapidly enough as $b \to \infty$.

2. Other Capacities: Open Questions. Another parameter associated with flow in fixed beds is the effective drag coefficient D. It is defined through the following problem:

$$\left.\begin{array}{l} \Delta u = \nabla P \\ \nabla \cdot u = 0 \end{array}\right\} \text{ in } Y^o(w) \tag{2.1, 2.2}$$

$$u = 0 \quad \text{on } \partial Y^o(\omega) \tag{2.3}$$

$$u \to \hat{e} \quad \text{as } |x| \to \infty . \tag{2.4}$$

We now let F be

$$F = \left\langle \int_{\partial Y^o(\omega)} \sigma(x,\omega) \cdot n \right\rangle, \tag{2.5}$$

where σ is the stress tensor. Then D is defined by

$$F = D\hat{e}. \tag{2.6}$$

Asymptotic expressions for D, when $Y^c(\omega)$ is an isotropic dilute distribution of spheres, were constructed by Childress [5], Hinch [8] and others. They found:

$$D = \left[1 + \frac{3}{\sqrt{2}}\,\phi_2^{1/2} + \frac{135}{64}\,\phi_2\log\phi_2 + O(\phi_2)\right]\frac{9}{2a^2}\phi_2 \tag{2.7}$$

Later people identified D^{-1} with the permeability K (see the similarity between (2.6) and (1.3)), and compared their bounds to (2.7). But are D^{-1} and K identical? To clarify the issue, let us consider the analogous problems in potential theory. We formulate the following two problems:

$$\begin{aligned} \Delta u &= 0 \quad \text{in } V \setminus B \\ u &= 0 \quad \text{on } \partial B \\ u &= 1 \quad \text{on } \partial V \end{aligned} \tag{2.8}$$

$$\begin{aligned} \Delta v &= -1 \quad \text{in } V \setminus B \\ v &= 0 \quad \text{on } \partial B \\ \frac{\partial v}{\partial n} &= 0 \quad \text{on } \partial V \ . \end{aligned} \tag{2.9}$$

Here V is a large domain in $\mathbb{R}^3$, and B is a collection of inclusions in V.

Next we define

$$\alpha_1 = \frac{1}{V}\int_{\partial V} \frac{\partial u}{\partial n}\ , \tag{2.10}$$

$$\alpha_2 = \left(\frac{1}{V}\int_{V\setminus B} v\right)^{-1} \tag{2.11}$$

α_1 is the classical capacity of $V \setminus B$, and we denote it here as boundary capacity. We call α_2 the interior capacity. In the problem of diffusion-controlled reactions, α_2 is called the rate constant [10]. While α_1 is the potential theory analog to the drag coefficient, α_2^{-1} is analogous to the permeability. It is easy to construct examples which show that they are different. When V and B are two concentric spheres, $\alpha_2 > \alpha_1$, and when V and B are two eccentric spheres with a small separation of $O(\epsilon)$ (see Figure 2) then $\alpha_1 \sim O(\epsilon^{-1}) \gg \alpha_2 \sim O(1)$. We remark that the dilute limits of D and K^{-1} are the same, that is, $\frac{9}{2a^2}\phi_2$. In fact, effective field approximations for K [3],[4] yield

$$K \sim K_s\left(1 - \frac{3}{\sqrt{2}}\,\phi_2^{1/2}\right) \text{ as } \phi_2 \to 0. \tag{2.12}$$

Thus K^{-1} and D seem to agree to second order. There is no reason to believe, however, that they are identical for arbitrary ϕ_2. Note that the bound (1.26) does not capture the nonanalytic (in ϕ_2) behavior predicted by (2.12). The reason is that the trial field used to obtain (1.26) does not incorporate screening effects.

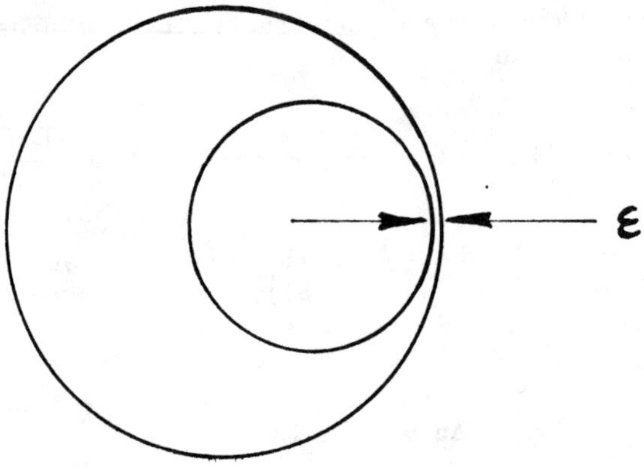

Figure 2: Two eccentric spheres

Finally, we say a few words about the behavior near percolation. When the void region $(Y^0(\omega))$ ceases to percolate, the solution to (1.4)-(1.5) is simply hydrostatic pressure, and zero velocity. Thus K = 0 at this point. This observation is related to the difficulties we had in obtaining lower bounds on K. The nearest neighbor distribution is local in nature, and it does not capture the long range correlations associated with percolation. As noted by Torquato and Rubinstein [10], the interior capacity α_2^{-1} (or the inverse rate constant) does not go to zero when the void region ceases to percolate. In this sense, permeability and rate constant problems are different.

REFERENCES

[1] J.D.Beasley and S.Torquato, *Phys.Fluids* A **1** (1989), 199.

[2] J.G. Berryman and G.W. Milton, *J. Chem. Phys. 83* (1985), 754.

[3] H.C. Brinkman, *Appl. Sci. Res. A1* (1947), 27.

[4] R. Caflisch and J. Rubinstein, *Lectures on the Mathematical Theory of Multiphase Flow*, Courant Institute Lecture Notes Series, 1986.

[5] S. Childress, *J. Phys. Soc. Jpn. 40* (1976), 567.

[7] J. Happel and H. Brenner, *Low Reynolds Number Hydrodynamics*, Nijhoff, 1983.

[8] E.J. Hinch, *J. Fl. Mech. 83*, (1977), 695.

[9] S. Prager, *Phys. Fluids 4* (1961), 1477.

[10] J. Rubinstein and S. Torquato, *J. Chem. Phys. 88* (1988), 6372.

[11] J. Rubinstein and S. Torquato, *J. Fl. Mech.* **206** (1989), 25.

[12] S. Torquato, *J. Stat. Phys.* 45 (1986),843.

[13] S. Torquato and J.D. Beasley, *Phys. Fluids 30* (1987), 633.

CHAPTER 5

Propagation of Singularities and Some Inverse Problems in Wave Propagation

William W. Symes*

Abstract. We review a number of results relating the propagation of singularities for hyperbolic partial differential equations — i.e. the persistence, or non-localization, of wave motion — with well-posedness for some inverse problems of reflection type, such as arise for instance in seismology and ultrasonics. By far the most complete information is available for layered problems. We show how a simple but refined propagation-of-singularities theorem, with estimates, yields important functional properties of the model-data relationship for such problems, including regularity in various useful coefficient classes, separation of scales, ... We explain the essential role of *travel time* in the study of these problems, and show how its function may be generalized to multidimensional (i.e. non-layered) problems.

1. Introduction. The "inverse problems in wave propagation" of the title are idealizations of remote sensing techniques such as reflection seismology, ground-penetrating radar, and pulse-echo ultrasonic nondestructive evaluation. A wave of some sort is stimulated by an artificial energy source. It propagates into a region containing the remote structure of interest; the changes in physical properties through the structure cause some of the energy of the wave to be diverted into a reflected wave or echo, which travels in roughly the opposite direction, back to a region where the experimentalist has installed appropriate sensors and recording devices. The records of reflected waves are to be analysed to reveal whatever can be inferred about the physical properties of the remote structure.

Two conditions must be fulfilled for this scheme to work:

1. Energy, propagating in the form of waves, must penetrate the region of the remote structure;

* Department of Mathematical Sciences, Rice University, Houston, TX.

2. A reflection or echo process must take place, partitioning the energy of the wavefield.

The requirement that energy reach the target area in the form of waves (and reciprocally, return from the target area as waves) rather than as a diffusion process, say, is motivated by a need for spatial resolution in applications. Evidently the resolution scale for a wave process might be measured in wavelengths, whereas the resolution possible with measurements of diffusing fields is much harder to understand.

Roughly speaking, wave processes are modelled by hyperbolic systems of partial differential equations. The association is justified by the *geometrical optics analysis* of high-frequency solutions, which has its modern expression in various *propagation of singularities* results. These theorems give precise statements to the effect that a disturbance which begins coherently along a wavefront will continue in that form, the wavefront moving in such a way that its orthogonal trajectories (in the simplest cases) obey a system of ordinary differential equations. The description of the disturbance thus obtained corresponds well to the physical notion of *wave*, and appears to be correct when the coefficients of the hyperbolic system, which parameterize local physics of the propagating medium, are smooth on the scale of the wavelength of the disturbance.

In particular, when these coefficients are smooth on a scale much longer than a wavelength, the geometric optics approximation predicts *no reflected waves at all.* This prediction is borne out quite well by numerical experiments. Accordingly, models of reflecting media must involve coefficients rough on the scale of a wavelength, or finer. Morlet (1982) has made an extensive numerical study of the reflection and transmission of *plane waves in plane layered acoustic media* (these will be described below) — that is, the coefficients of the acoustic wave equation are functions of one variable only. Such plane-layered models are widely used in seismology, as for various reasons the mechanical properties of rocks often vary mostly in the vertical direction. Morlet found that once the length scale of strong variation in coefficients decreased to the level of a wavelength *and* the propagation distance remained constant, the wave nature of the propagating disturbance was strongly modified, and below some scale disappeared altogether: the wave did not propagate through such strongly oscillatory media at all. Since no "energy" penetrated the highly oscillatory region, there were no reflected waves out of which to infer its structure. That is, the wave was *localized* in the complement of the oscillatory region.

Thus models of media which are *too smooth* on the wavelength scale do not generate reflected waves because all of the "energy" in the wavefield passes through unaltered. Models which are *too rough*, on the other hand, generate no reflected waves because no "energy" penetrates to be reflected. ("Energy" here merely means some convenient measure of the local size of the field, which may or may not be related to a physical energy.) Thus the questions are naturally posed:

1. Do there exist models of a degree of roughness/smoothness *just right* to permit wave propagation *and* generate sufficiently informative reflected waves?

2. Do such models, if they exist, model real physical systems?

The first question is mathematical; partial answers will be sketched in the following pages. The second question does not have definitive answers, so far as the author is aware, though the motivation for question (1) is the *appearance* of experimental data from a variety of subjects. Clearer answers to question (2) await a combination of improvements in theory, numerical algorithms, and experimental technique.

While the generalities should be similar for the various physical models supporting wave propagation (electromagnetism, linear elastodynamics, ...), the details are best understood for linear acoustics, governed by the linearized pressure equation

$$\frac{1}{\rho c^2}\frac{\partial^2 u}{\partial t^2} - \nabla \cdot \frac{1}{\rho}\nabla u = F \text{ in } \mathbb{R}^n, \quad n = 1, 2, 3, \ldots$$

where $u(x,t)$ is the (scalar, infinitesimal excess) pressure field in a fluid, $\rho(x)$ is the density in the equilibrium state, $c(x)$ is the sound velocity, and the source term F is the divergence of a body force. We shall consider both

(i) F vanishes, and u has the form of a *plane wave*

$$u(x,t) = U(t - \theta \cdot x)$$

for large negative t — sensible when ρ and c are constant in a half-space, say (§§2–4);

(ii) u vanishes identically for large negative t, and F has point support, i.e.

$$F(x,t) = f(t)\delta(x - x_s) \qquad (\S5)$$

The data of an inverse problem of the type considered here is the trace of the pressure field u on a time-like hypersurface — we shall consider only a coordinate hyperplane:

$$S[\rho, c; F] = u|_{x_n=0} \cdot$$

Of course, in reality, measured signals are *sampled*, both spatially and temporally, and also *filtered* by the recording apparatus. In this paper we ignore such refinements.

Since the local "physics" of this model are determined entirely by ρ and c, the object of the inverse problem is to determine (to the extent possible) $\rho(x)$ and $c(x)$ for $x_n \neq 0$ from $S[\rho, c; F]$. (F will be regarded as known, though in fact it should be included in the unknowns for many models).

By far the best results are available for the plane-wave problem for layered media, i.e. $\rho = \rho(x_n)$, $c = c(x_n)$. In Section 2 we state a simple but quite precise propagation-of-singularities result for such problems, and exhibit the plethora of consequences which follow from it: regularity properties of the map S, rough correspondence between scales in u and scales in ρ, c, etc. We describe key results in the simplest form, and refer the reader to the literature for the most general versions. In §3 we show how these consequences of a refined propagation-of-singularity result can be used to investigate the plane-wave-inverse problem. The end result is a well-posedness result for "mildly rough" models: thus we are able to give an affirmative answer to question (1) above in this special instance. A very important role is played in this development by the reparameterization of the layered model by travel-time: *without this device, essentially nothing can be accomplished, either theoretically or numerically.* In Section 4, we give a very brief treatment of rougher media, for which a dichotomy of scales exists, so that homogenization estimates can be used to reduce the problem to the mildly rough case.

The plane-wave/layered medium problems, while forming an excellent arena for development of insight, offer limited opportunity for immediate practical application. We consider the general multidimensional "point-source" problem (case (ii)) in §5. Very little is known about this class of problems, relative to the layered case. We address two issues:

(1) proper definition of the map S on non-smooth coefficient classes;

(2) location of a substitute for the travel-time transformation.

For both issues, we can offer only fragmentary information. We are able, however, to explain the function of the travel-time transformation for the plane-layered problems in such a way that its functional substitute for the non-layered problem is (essentially) uniquely specified. We give a construction of this "multidimensional travel-time transformation" under very restrictive smoothness assumptions, which (despite the restrictions) indicates that such things are actually possible — this is the only original result of the paper.

Most of the results and insights reported here are the result of joint work with many colleagues insofar as they are due to the author at all. Especially Fadil Santosa and Paul Sacks deserve much credit for whatever is worthwhile in the following pages. Special thanks are also due to Rakesh, R.M. Lewis, Cheryl Percell, and Gang Bao.

2. Some Simple Propagation-of-Singularities Results in Layered Media, and Consequences. When the density ρ and sound velocity c are smooth and functions of only one space coordinate, say z, then the acoustic wave equation

$$\left(\frac{1}{\rho c^2}\frac{\partial^2}{\partial t^2} - \nabla\cdot\frac{1}{\rho}\nabla\right)u = 0 \tag{2.1}$$

has *plane-wave solutions*, i.e. the pressure $u(x,z,t)$ is in the form

$$u(x,z,t) = U(z, t-\xi\cdot x)$$

Such solutions exist for $\xi \in \mathbb{R}^n$ so long as $c(z)|\xi| < 1$ for all z. Evidently U is a solution of

$$\left[\frac{1}{\rho}\left(\frac{1}{c^2} - |\xi|^2\right)\frac{\partial^2}{\partial t^2} - \frac{\partial}{\partial z}\frac{1}{\rho}\frac{\partial}{\partial z}\right]U = 0\,. \tag{2.2}$$

Write $v = (c^{-2} - |\xi|^2)^{-\frac{1}{2}}$ and suppress ξ from the discussion for the remainder of this section. If $v \equiv v_0 \equiv$ const. for $z < 0$, and $a_0 \in \mathbb{R}$, then there exists a unique distribution solution to (2.2) with

$$U(z,t) = a_0 H(t - z/v_0)$$

for $t, z < 0$. Standard constructions, e.g. Courant and Hilbert (1962, Ch. V) imply that U has the progressing wave expansion

$$U(z,t) = a(z,t)H(t - \tau(z))$$

where

$$\tau(z) = \int_0^z \frac{1}{v}$$

is the *travel time*.

To pursue the agenda outlined in §1, one needs explicit estimates of a in terms of norms of ρ, c. Also, it will be advantageous to consider the slightly more general problem

$$\begin{aligned} \frac{1}{\rho v^2}\frac{\partial^2 U}{\partial t^2} - \frac{\partial}{\partial z}\left(\frac{1}{\rho}\frac{\partial U}{\partial z}\right) &= \psi_0(z)\delta(t-\tau(z)) + \psi_1(z)w(z,t)H(t-\tau(z)) \\ U &= 0, \qquad t << 0 \end{aligned}$$

with ψ_0, ψ_1 and w smooth. A simple calculation shows that we should take $U = aH(t - \tau)$ with a the solution of the characteristic Cauchy problem

$$\begin{aligned} \frac{1}{\rho v^2}\frac{\partial^2 a}{\partial t^2} - \frac{\partial}{\partial z}\left(\frac{1}{\rho}\frac{\partial a}{\partial z}\right) &= \psi_1 w H(t-\tau) \quad \text{for } t > \tau(z)\,; \\ \frac{d}{dz}a(z,\tau(z)) - \sigma(z)a(z) &= \rho(z)v(z)\psi_0(z), \quad z > 0 \\ a(0,0) &= a_0 \end{aligned} \tag{2.3}$$

where $\sigma(z) = \frac{d}{dz}(\log \rho v)(z))$. Following arguments in Symes (1986a; Section 1), we obtain estimates for the "sideways energy form"

$$Q_T(z) = \frac{1}{2}\int_{\tau(z)}^{2T-\tau(z)} dt \left(\left|\frac{\partial U}{\partial t}\right|^2 + \left|\frac{\partial U}{\partial z}\right|^2\right)$$

of the form:

$$\begin{aligned} &\|Q_T\|_{L^\infty[0,Z(Y)]} \leq \\ &\quad C\left(\|\psi_0\|^2_{L^2[0,Z(T)]} + a_0^2\|\sigma\|^2_{L^2[0,Z(T)]}\right. \\ &\quad \left. + \|\psi_1\|^2_{L^2[0,Z(T)]} \sup_{0\leq z\leq Z(T)}\left\{\int_{\tau(z)}^{2T-\tau(z)} dt|w(z,t)|^2\right\}\right) \end{aligned} \tag{2.4}$$

for suitable $C > 0$ depending *only* on $\|\rho\|_{H^1[0,Z(T)]}$, $\|v\|_{H^1[0,Z(T)]}$. Here

$$Z(T) = \inf\{z : \tau(z) \geq T\}\,.$$

Since we also have control of the characteristic boundary condition $a(z, \tau(z))$ through (2.3), an identical bound follows for the H^1 norm of U in $L(T) = \{(z,t) : 0 \leq z \leq Z(T),\ \tau(z) \leq t \leq 2T - \tau(z)$. On the other hand, (2.4) is stronger than such a statement, and shows that the traces of first derivatives of U on the vertical line segments $\{z = \text{const.}\}$ are well defined.

Thus (2.4) is a rather strong propagation-of-regularity result — probably the strongest that can be expected to hold uniformly over ρ, v in $C^\infty \cap$ a bounded set in H^1_{loc}. It immediately allows us to extend the definition of various maps beyond the class of smooth coefficients. For example, suppose that $a_0 = 1$, ρ_1, ρ_2 both lie in the same bounded set in $H^1_{\text{loc}} \cap C^\infty$, and $v \equiv 1$. The difference $V = U_2 - U_1$ satisfies

$$\frac{\partial^2 V}{\partial z^2} - \frac{\partial^2 V}{\partial z^2} + \sigma_1 \frac{\partial V}{\partial z} = (\sigma_2 - \sigma_1)\frac{\partial U_2}{\partial z}$$

where $\sigma_i = \frac{d}{dz}\log\rho_i$, $i = 1, 2$. From (2.4) with $\psi_0 = \psi_1 = w \equiv 0$ and $a_0 = 1$, we get a bound

$$\begin{aligned}\int_{\tau(z)}^{2T-\tau(z)} dt \left|\frac{\partial U_2}{\partial z}\right|^2 &\le Q(z) \le C\|\sigma_2\|^2_{L^2[0,Z(T)]} \\ &\le C\|\rho_2\|_{H^1[0,Z(T)]}\end{aligned}$$

so we can use (2.4) again with $\psi_0 \equiv 0$, $\psi_1 = \sigma_2 - \sigma_1$, $w = \partial U_2/\partial z$, and $a_0 = 0$ to get

$$\int_{\tau(z)}^{2T-\tau(z)} dt \left(\left|\frac{\partial V}{\partial t}\right|^2 + \left|\frac{\partial V}{\partial x}\right|^2\right) \le C\|\sigma_2 - \sigma_1\|^2_{L^2[0,Z(T)]}$$

whence a Lipschitz estimate follows for the map

$$\tilde{S}_\delta : \log\rho \in H^1[0, Z(T)] \mapsto \frac{\partial U}{\partial t}(0, \cdot) \in L^2[0, 2T]$$

for example, uniform over bounded sets in $H^1[0, Z(T)] \cap C^\infty(\mathbb{R})$. Consequently the map extends — in some sense, we have "solved" this singular problem with non-smooth (H^1) coefficients.

In Symes (1986a) these arguments are developed further to show that $\tilde{S}_\delta$ is actually a C^1 diffeomorphism and further work along the same lines shows that $\tilde{S}_\delta$ is C^2. It is probably C^∞. Since $\tilde{S}_\delta$ is differentiable, estimates for the derivative are useful. It is possible to show, by very similar arguments, that the formal linearization

$$\left(\frac{1}{\rho v^2}\frac{\partial^2}{\partial t^2} - \frac{\partial}{\partial z}\frac{1}{\rho}\frac{\partial}{\partial z}\right)\delta U = -\frac{\partial \sigma}{\partial z}\frac{1}{\rho}\frac{\partial U}{\partial z} \qquad \delta U \equiv 0\,,\ \ t << 0$$

actually defines the derivative

$$D\tilde{S}_\delta[\rho]\delta\rho = \frac{\partial \delta U}{\partial t}(0, \cdot)\,.$$

Moreover,

$$C_*\|\delta\rho\|_{H^1[0,Z(T)]} \le \|D\tilde{S}_\delta[\rho]\delta\rho\|_{L^2[0,Z(T)]} \le C^*\|\delta\rho\|_{H^1[0,Z(T)]}$$

where the constants $0 < C_* \le C^*$ are uniform as $\log\rho$ ranges over bounded sets in H^1_{loc}. Similarly,

$$D\tilde{S}_H[\rho]\delta\rho = \delta U(0, \cdot)$$

satisfies

$$\|D\tilde{S}_H[\rho]\delta\rho\|_{L^2[0,2T]} \le C_H \|\delta\rho\|_{L^2[0,Z(T)]}$$

with another H^1_{loc}-uniform constant C_H, though no lower estimate seems to obtain for $D\tilde{S}_H$.

These estimates show that there is *approximate separation of scales* in the relation between coefficients and solution, in the following sense. A convolution kernel $f \in \mathcal{E}'$ is *elliptic* on H^1_{loc} if there exist constants K_0, K_1, K^* so that for $u \in H^1(\mathbb{R})$,

$$K_1\|u\|_1 - K_0\|u\|_0 \le \|f * u\|_1 \le K^*\|u\|_1 \,.$$

For example, if $|\hat{f}(\omega)| \ge p$ for $|\omega| \ge \Omega_*$, then one could take crudely $K_1 = p, K_0 = (1+\Omega_*^2)^{\frac{1}{2}}p$ (much more precise statements are possible). Thus the relation between the constants K_0, K_1 measures the threshold of the "passband" of f. Such f may be manufactured by subtracting from the Dirac δ function a smooth approximation. Without loss of generality we may assume that f is "causal", i.e. supp $f \subset [0,\infty)$.

As a consequence of the above estimates for $D\tilde{S}_\delta$, the derivative of

$$\tilde{S}_f = f * \tilde{S}_\delta$$

satisfies

$$\begin{aligned} \|D\tilde{S}_f[\rho]\delta\rho\| \;\ge\; & \tilde{K}_1\|\delta\rho\|_{H^1[0,Z(T)]} - \tilde{K}_0\|\delta\rho\|_{L^2[0,Z(T)+Z_f]} \\ & - \tilde{K}_m m_1(f)\|\delta\rho\|_{H^1[Z(T),Z(T)+Z_f]} \end{aligned} \tag{2.5}$$

where $\tilde{K}_1, \tilde{K}_0$ and $\tilde{K}_m$ depend on $\|\log\rho\|_{H^1[0,Z(T)]}$, the constants K^*, K_1 and K_0 in the ellipticity estimate of f, and K_m and Z_f depend also on the *first moment*

$$m_1(f) = \inf_{t_0\in\mathbb{R}} \sup_{\phi\in C_0^\infty(\mathbb{R})} \frac{\langle f,(t-t_0)\phi\rangle}{\|\phi\|_{L^\infty}}$$

of f.

The first two terms in the preceding estimate show that $D\tilde{S}_f$ has essentially the same "filter" or separation-of-scales effect as does $f*$, at least qualitatively. The last term is necessary because the "breadth" of f, as measured by its first moment $m_1(f)$, may be nonzero, so that control of $\delta\tilde{S}_f$ is needed over a slightly longer time interval — or equivalently of $\delta\rho$ over an "extra" depth interval — to ensure control over $\delta\rho$ on $[0, Z(T)]$. Both the coefficient $K_m m_1(f)$ and the size Z_f of the extra depth interval are $O(m_1(f))$.

In order to deal with truly "bandlimited" kernels — i.e. $f \in C_0^\infty$ as well — higher-order estimates are needed. Suzuki (1988) showed (essentially) that $\tilde{S}_{\delta'} = \partial/\partial t\tilde{S}_\delta$ obeys an estimate

$$\|D\tilde{S}_{\delta'}[\rho]\delta\rho\|_{L^2[0,2T]} \le C_2\|\delta\rho\|_{H^2[0,Z(T)]}$$

using extensions of the arguments sketched above; the constant is uniform over bounded sets in H^2_{loc}. This estimate can be used to study $\tilde{S}_f$ when $f*$ obeys an estimate of the form

$$K_1\|u\|_1 - K_0\|u\|_0 - K_2\|u\|_2 \le \|f * u\|_1 \le K^*\|u\|_1 \,. \tag{2.6}$$

Once again, this estimate can be interpreted loosely as identifying the "passband" of f. If $|\hat{f}(\omega)| \geq p > 0$ for $\Omega_* \leq |\omega| \leq \Omega^*$, one can take $K_1 = p$, $K_0 = p(1+\Omega_*)^{\frac{1}{2}}$ as before and $K_2 = p(1+\Omega^*)^{-\frac{1}{2}}$. There results for $D\tilde{S}_f$ an estimate of the form

$$\begin{aligned} \|D\tilde{S}_f[\rho]\delta\rho\|_{L^2[0,2T]} \geq\ & \tilde{K}_1\|\delta\rho\|_{H^1[0,Z(T)]} - \tilde{K}_0\|\delta\rho\|_{L^2[0,Z(T)]} \\ & - \tilde{K}_2\|\delta\rho\|_{H^2[0,Z(T)]} \\ & - \tilde{K}_m m_1(f)\|\delta\rho\|_{H^1[Z(T),Z(T)+Z_f]} \end{aligned}$$

all dependences being as before, except for $\tilde{K}_2$ which is

(a) uniform over bounded sets in H^2_{loc} for fixed f;

(b) for ρ in a fixed bounded set in H^2_{loc}, $\tilde{K}_2 = O(K_2)$.

Thus the estimate is nontrivial if $\tilde{K}_2$ is small enough, i.e. if f is "close enough to elliptic" relative to $\|\rho\|_{H^1[0,Z(T)]}$.

To obtain similar results for the map $v \mapsto U$, or $(\rho, v) \mapsto U$, one must impose much more smoothness on v, essentially because a change in v actually moves the characteristics. See e.g. Lewis's thesis (1989) or Symes (1986b). The diffeomorphism is lost: in metrics strong enough to make $v \mapsto U$ differentiable, the derivative is not coercive.

3. Travel Time and the Determination of "Mildly Rough" Coefficients. Maintain the setting of the previous section, but now consider, for simplicity, the case $\rho \equiv$ const. $(= 1)$ instead. Introduce the (*travel-time*) change-of-variable

$$x = \tau(z)$$

(NOTE: we are using "x" in a different sense than at the beginning of §2!) Denoting by ζ the inverse function of τ, so that

$$x = \int_0^{\zeta(x)} \frac{1}{v}$$

introduce

$$\tilde{v}(x) = v(\zeta(x)), \quad \tilde{U}(x,t) = U(\zeta(x),t)\,.$$

A short calculation verifies that

$$\tilde{U}(x,t) = H(t-x), \qquad x,t < 0\,;$$

$$\frac{\partial^2 \tilde{U}}{\partial t^2} - \frac{\partial^2 \tilde{U}}{\partial x^2} - \tilde{\sigma}\frac{\partial \tilde{U}}{\partial x} = 0$$

where $\tilde{\sigma} = \frac{d}{dx}\log \tilde{v}$. Thus $\tilde{v}$ plays exactly the same role as ρ did at the end of the last section, hence the map

$$\tilde{S}_\delta : \tilde{v} \in H^1[0,T] \to \frac{\partial \tilde{U}}{\partial t}(0,\cdot) \in L^2[0,2T]$$

is well-defined by continuous extension from $\tilde{v} \in C^\infty$, is a C^2 diffeomorphism, and all the "quasi-elliptic" estimates in §2 apply ipso-facto to $\tilde{S}_f = f * \tilde{S}_\delta$.

Remark In §5 some commentary is offered on the "real" significance of this change of variable.

To understand how results about $\tilde{S}_f$ might yield information about v (rather than $\tilde{v}$), examine the change-of variables operator:

$$M[v] = v \circ \zeta\,.$$

It is possible to show that

(i) M extends to a continuous map defined on the positive cone

$$\Gamma_k = \{v \in H^k_{\text{loc}}[0,\infty) : v(0) = v_0, v(z) > 0\ \forall z\}$$

and taking values in Γ_k, for $k \geq 1$;

(ii) M is nowhere locally uniformly continuous, hence *a fortiori* not differentiable, in Γ_k.

(iii) Regarded as a map: $\Gamma_{k+2} \to \Gamma_k$, M is differentiable.

For discussion and proofs of related results see Lewis (1989) and Symes (1986b). The continuity result is due to Rakesh (unpublished).

Formally, the derivative of M is given by

$$\begin{aligned}\delta M(t) &= \delta v(\zeta(t)) + v'(\zeta(t))\delta\zeta(t)\\ &= \left(\delta v + vv'\int_0^z \frac{\delta v}{v^2}\right)\Bigg|_{z=\zeta(t)}.\end{aligned}$$

Thus the "principal part" of δM is the change-of-variables for v, applied to δv — the simplest sort of Fourier integral operator. In fact, if we assume that $v =$ const. for large Z, then

$$\frac{d}{dz}M[v] = \int dk\ a(t,k)e^{i\zeta(t)k}\left(\frac{dv}{dz}\right)^\wedge(k)$$

where the phase ζ itself depends on v — i.e. dM/dz is a "nonlinear Fourier integral operator."

On the other hand, the appearance of v' in the second term above hints at the nondifferentiability of $M : \Gamma_k \to \Gamma_k$.

Now suppose that f is an elliptic convolution kernel, as in the previous section. Appraisal of the "Gårding" estimate (2.5) shows that, if $\delta\tilde{v}$ is sufficiently smooth — that is, $\|\delta\tilde{v}\|_{L^2}$ is sufficiently big relative to $\|\delta\tilde{v}\|_{H^1}$ — the estimate has no force. If $\delta\tilde{v} = \delta M$, with smooth δv, however, it may be that δv (rather than $\delta\tilde{v}$) is controlled. In fact, if $\|v'\|_{H^1}$ is large, one might suspect that the second term in the expression for δM dominates $\|\int \frac{\delta v}{v^2}\|_{L^2}$, hence (for smooth δv) $\|\delta v\|_{L^2}$ itself. This idea is embodied in the "rough/smooth" lemma (see Symes (1988), Lemma 3):

For $u, \Phi \in C^\infty$, $a < b$, $\Delta > 0$, *set*

$$r(x,\Delta) = \frac{1}{\Delta}\int_{x-\frac{\Delta}{2}}^{x+\frac{\Delta}{2}} |u|^2$$

$$r_*(\Delta) = \inf_{x\in[a,b]} r(x,\Delta), \quad r^* = \sup_{x\in[a,b]} r(x,\Delta)$$

Then

$$\begin{aligned} \|\Phi u\|^2_{L^2[a,b]} \ \geq\ & \frac{r_*(\Delta)}{2}\|\Phi\|^2_{L^2[a,b]} \\ & - \frac{16}{9}(r_*(\Delta)+r^*(\Delta))\Delta^2\|\Phi'\|^2_{L^2[a,b]} \end{aligned} \tag{3.7}$$

Apply this estimate to the second term in δM: with $u = \left(\frac{1}{2}v^2\right)''$, $\Phi = \int \cdot \frac{\delta v}{v^2}$, we get

$$\begin{aligned} \left\|\left(\delta v + vv'\int\frac{\delta v}{v^2}\right)\circ\zeta\right\|^2_{H^1[0,T]} \ \geq\ & C_1 r_*(\Delta)\left\|\int\frac{\delta v}{v^2}\right\|^2_{L^2[0,Z(T)]} - C_2\|\delta v\|^2_{H^1[0,Z(T)]} \\ & - C_3(r^*(\Delta)\Delta^2+1)\,\|\delta v\|^2_{L^2[0,Z(T)]} \end{aligned}$$

where C_1, C_2, and C_3 are uniform for v in bounded sets in H^1_{loc}. Thus if δv is smooth — so that $\|\delta v\|_{H^1}$, $\|\delta v\|_{L^2}$ and $\|\int\delta v\|_{L^2}$ are comparable — *and* if v'' is uniformly big relative to $\|v\|_{H^1}$, locally on a length scale of Δ, *and* Δ is small enough, — then

$$\|\delta M\|_{H^1[0,T]} \geq C\|\delta v\|_{L^2[0,Z(T)]}\,. \tag{3.8}$$

Define $S_f = \tilde{S}_f \circ M$. Combining (3.8) with (2.5) we see that if in addition $\|\delta v\|_{L^2}$ is big enough relative to $\|\delta v\|_{H^1}$, then

$$\|\delta S\|_{L^2[0,2T]} \geq C\|\delta v\|_{L^2[0,Z(T)]}$$

(presuming $\delta v \equiv 0$ in $(Z[T], Z[T]+Z_f)$ in order not to have to worry about the "margin correction"). That is, δS is coercive, in a certain sense, for very smooth δv. On the other hand it is plain that δS is coercive for oscillatory δv.

Unfortunately, that's as far as the purely 1-D problem goes. If δv is neither purely smooth nor purely oscillatory, it is possible to arrange for the two parts of δM to very nearly cancel, even when v satisfies the "uniform roughness" condition sketched above, so that δS cannot possibly be coercive.

The way out is to remember — from the beginning of §2 — that each choice of sound speed profile $c(z)$ (keeping ρ constant still) gives a *suite* of one-dimensional problems with plane-wave velocities

$$v(z,\xi) = c(z)(1-c^2(z)\xi^2)^{-\frac{1}{2}}\,.$$

The change-of-variables operator M and the seismogram operators S_f thereby acquire dependence on ξ as well, of course. The perturbation in v is related to the perturbation in c by

$$\delta v = \delta c(1-c^2\xi^2)^{-\frac{3}{2}}$$

so

$$\delta M = (1-c^2\xi^2)^{-\frac{3}{2}} \left[\delta c - cc'(1-c^2\xi^2)^{-\frac{1}{2}} \int^z \frac{\delta c}{c}(1-c^2\xi^2)^{-\frac{1}{2}} \right] \Bigg|_{z=\xi} .$$

It is not surprising that, while δM can be made to vanish at *one* value of ξ, it *cannot* be made to vanish over a range of ξ simultaneously. In fact, a coercivity estimate is possible of the form

$$\|\delta M\|_{H^1([0,T]\times[\xi_{\min},\xi_{\max}])} \geq C\|\delta c\|_{L^2[0,Z(T)]}$$

when C is "rough enough."

The preceding is only the barest sketch of an argument, of course. In Symes (1988) we have shown that all of the choices mentioned above can actually be made. The end result is the identification of a class $\Sigma \subset H^1_{\text{loc}}$ of velocity profiles c and of suitable "elliptic" kernels f which permit stable solution of the inverse problem. More precisely, for each elliptic kernel f, there exists a neighborhood V_f of $S_f(\Sigma)$ in L^2_{loc} and a locally Lipschitz map $I_f : V_f \to L^2_{\text{loc}}$ satisfying

$$I_f(S_f(c)) = c \quad \text{for} \quad c \in \Sigma .$$

When f is "bandlimited," i.e. smooth but obeying a quasi-elliptic estimate like (2.6), a similar result is obtained *via* regularization: if the "upper bandlimit" parameter K_2 is small enough, a stable inverse map I_f exists for which $[I_f(S_f(c)) - c]$ is small — as usual with regularization, one does not obtain exact recovery of the model. The degree of error (for consistent data in $S_f(\Sigma)$) can be made as small as one likes, at the usual price of degradation of stability.

The construction of I depends on a variational problem of least-squares type. This variational problem has served as the basis of a computer program to solve the inverse problem numerically, which has yielded quite promising results when applied to both synthetic and field seismic data. For details see Symes (1988), Carazzone and Symes (1989), and references cited there.

4. Very Rough Models and Homogenization. The sedimentary crust, amongst other real materials, does not appear to possess the smoothness relative to wavelengths and other scales necessary to make plausible direct application of the results of the last section. Instead, very rough models may be *approximated* in a weak sense by smooth ones, provided that a dichotomy of length scales is present.

For 1-dimensional wave propagation, a version of such an approximation scheme may be based on the quantitative homogenization estimates developed by Bamberger *et al.* (1979). Specialized to the present context, this estimate amounts to

$$\|S_f[c_1] - S_f[c_2]\|_{L^2[0,2T]} \leq C\|f\|_{H^4(\mathbb{R})}\Gamma(c_1, c_2)$$

for $\log c_1, \log c_2 \in L^\infty_{\text{loc}(\mathbb{R})}$. The pseudo-distance Γ is defined by

$$\Gamma(c_1, c_2) = \sup_{0\leq z\leq Z} \left| \int_0^z dz' \left(\frac{1}{c_1^2(z')} - \frac{1}{c_2^2(z')} \right) \right|$$

where Z is chosen sufficiently large.

If Σ is the set of models described in the previous section, V_f the neighborhood of $S_f(\Sigma)$, and f is a "quasi-elliptic" but smooth kernel, we can associate to the pair (Σ, f) the set Σ_f of $\log c \in L^\infty_{\text{loc}}$ for which $S_f[c] \in V_f$. It is evident from the definition that Σ_f contains quite a bit more than an L^2-neighborhood of Σ. An easy way to see this is to pick a Dirac kernel $\phi \in C_0^\infty(\mathbb{R})$, $\phi(0) \neq 0$, $\int \phi = 1$, and set for $\epsilon > 0$ $\phi_\epsilon(z) = \epsilon^{-1}\phi(\epsilon^{-1}z)$. Roughly, ϵ is the "width" of ϕ_ϵ, and the convolution with ϕ_ϵ *averages* on the length scale ϵ. Then for $\log c \in L^\infty_{\text{loc}}$

$$\Gamma(c, \phi_\epsilon * c^{-2})^{-\frac{1}{2}} = O(\epsilon) \, .$$

Thus $c \in \Sigma_f$ if $\phi_\epsilon * c \in \Sigma$ for ϵ small enough that $S_f[c] \in V_f$. Such c can have arbitrarily large variation, provided that it is on a sufficiently fine length scale that averaging over length scale $\epsilon = O(\|f\|^{-1}_{H^4})$ yields an element of Σ. Σ is characterized also by being a subset of a large ball in H^2, the radius being determined by the "upper bandlimit" of the kernel (i.e. K_2 in 2.6). Thus the scales permitted in Σ are determined by this upper bandlimit. All significant components at shorter scales must be removed (essentially) by averaging on the length scale ϵ. Thus the difference $c - (\phi_\epsilon * c^{-2})^{-\frac{1}{2}}$ must be oscillatory on a scale smaller than ϵ, and much smaller than that permitted by membership in Σ.

These choices of scale can be made precise. In an unpublished technical report, the author made most of these choices explicit and investigated the question numerically. The results lent credence to the heuristic rule:

> *For effective recovery of the layered medium, the velocity profile must consist of a "core" medium which oscillates on a wavelength scale, and perhaps a very rough component oscillating on a subwavelength scale.*

Such a clean frequency dichotomy is probably not seen in nature; in fact some evidence exists that actual velocity distributions are *fractal.* Thus oscillatory components exist at all scales, and the preceding heuristic condition is not satisfied. It is not clear whether the "mildly rough" results bring with them results about media without the frequency dichotomy described here.

Papanicolau, White, and colleagues have studied another class of media admitting a dichotomy of length scales (Burridge *et al.* (1989)). The principal difference in problem setting between our work and theirs is in the relation between the source frequency content and the length scale of oscillation in the coefficients. Our work pertains to a fixed source, hence fixed bandwidth, and the limit as the length scale for the coefficients goes to zero. White *et al.* consider instead a family of sources with frequency increasing roughly as the square-root of the frequency of oscillation in the parameters (after normalization to fixed propagation time). Thus in their work the oscillations in the coefficient continue to interact with the oscillatory energy in the solution, which does not approach that of a smooth-coefficient equation. It is not clear which asymptotic regime is more important in explaining practical phenomena. Most likely both regimes are present in real data.

5. Multidimensional Problems. We return to the problem specified at the beginning of Section 2, viz. linear acoustics, but this time in dimension $n > 1$ and

with the coefficients allowed to depend on all the space variables. We introduce "energy" into the system *via* a "point source," i.e. a right-hand side with point support in x:

$$\left(\frac{1}{\rho(x)c^2(x)}\frac{\partial^2}{\partial t^2} - \nabla\cdot\frac{1}{\rho(x)}\nabla\right)u(x,t) = f(t)\delta(x) \qquad u\equiv 0,\ t<0\,.$$

Once again we are interested in estimates which hold for elliptic f, uniformly over bounded sets of coefficients in Sobolev spaces and especially in estimates for timelike traces.

Very little is known about these problems, and the following discussion is meant mainly to illustrate some of the remaining difficulties. Some partial results may be found in Rakesh (1988), Sacks and Symes (1985), and Sun (1987). Two major differences emerge compared to the layered case. The first of these is that considerably more coefficient regularity is needed to assure the optimal regularity of timelike traces. To fix ideas, take $f=\delta^{(-\frac{n-1}{2})}$ above, and assume that $\log c\in C^\infty$ is fixed. Near $x=t=0$, Hadamard's construction (Courant and Hilbert, Ch. 6) yields

$$u(x,t) = a(x)\rho^{-\frac{1}{2}}(x)H(t-\tau(x)) + R_0(x,t)$$

where $\tau(x)$ is the geodesic distance from x to 0 in the metric $c^{-2}\sum_i dx_i dx_i$, a is a C^∞ transport coefficient, and R_o vanishes at $t=\tau(x)$. This expression remains correct so long as we encounter no point conjugate to 0, i.e. as long as no caustics appear in the wave-front, which we assume (the complication in this analysis arising from caustics have not been understood, nor has the effect of changing c — see some discussion below). Thus u may as well be viewed as the solution of an interior characteristic Cauchy problem.

It is natural to attempt to bound the *energy norm*: With

$$\begin{aligned} B_T &= \{x : \tau(x)\le T\} \\ C_T &= \{(x,t) : t=\tau(x)\le T\} \end{aligned}$$

we get the standard energy identity

$$\int_{B_T} dx\left[\frac{1}{\rho c^2}\left|\frac{\partial u}{\partial t}\right|^2 + \frac{1}{\rho}|\nabla u|^2\right] = \int_{C_T}\frac{1}{\rho}\left|\frac{\partial u}{\partial t}\nabla\tau + \nabla u\right|^2 .$$

The transport equation (Hadamard's construction, above) yields all of the derivatives of u *tangential* to C_T, but the normal derivative is also needed to evaluate the R.H.S. of the energy identity. The normal derivative — or equally well, the t-derivative — trace on C_T is determined by the second transport equation,

$$\begin{aligned} &\nabla\tau\cdot\nabla a_1 + (b+\nabla\log\rho\cdot\nabla\tau)a_1 = \\ &(\Delta - \nabla\log\rho\cdot\nabla)(a\rho^{-\frac{1}{2}}) \\ &\lim_{t\to\tau(x)+}\frac{\partial u}{\partial t}(x,t) = a_1(x)\,. \end{aligned}$$

Here b depends on the bicharacteristic geometry, hence on c. Thus second derivatives of ρ are involved, and will not be "integrated out," an important difference with the layered case.

A certain amount of evidence suggests that the proper way to resolve this and other problems is to impose *microlocal* restrictions on the coefficients. Fundamentally, all of this evidence rests on Rauch's Lemma (Rauch, 1979), which we state here in a lopsided form suitable for application to linear problems, using now-conventional notation (Beals and Reed, 1984):

> *Suppose that for some $(x_0,\xi_0) \in T^*\mathbb{R}^n$, the distributions a, u satisfy $a \in H^s \cap H^r_{m\ell}(x_0,\xi_0)$, $u \in H^t \cap H^q_{m\ell}(x_0,\xi_0)$ with $0 \le t \le s$, $q \le t+r-s$, $q < s+t-n/2$. Then $a\,u \in H^t \cap H^q_{m\ell}(x_0,\xi_0)$.*

That is, $H^t \cap H^q_{m\ell}(x_0,\xi_0)$ is an $(H^s \cap H^r_{m\ell}(x_0,\xi_0))$–algebra. The lemma is usually stated for $t = s$, $q = r$, as is appropriate for applications to nonlinear problems. For linear problems such as that considered here, the solution may well have a different degree of smoothness than the coefficients — in fact, the solution u above is expected to have a jump discontinuity. Such strong singularities are much more difficult to treat in nonlinear problems, and most of the results based on Rauch's lemma and related ideas concern much weaker singularities — see e.g. Beals and Reed (1982, 1984). Nonetheless, we expect to be able to understand the relation between the coefficient and solution regularity *away* from the leading singularity or wavefront. For example, a quantified version of Rauch's lemma enables one to prove microlocal energy estimates for first-order hyperbolics like the transport equation above. In the present instance, we obtain $a_1 \in H^s$ if $\rho \in H^{s+2}$, $s > 1+n/2$ (n = number of *space* dimensions) plus microlocal conditions. Then Theorem 2.1 in Beals and Reed (1984) implies that $a_1 \in H^{s+1}_{m\ell}$ away from the characteristic variety of the transport operator. One then iterates this construction to produce higher energy estimates for u interior to the light cone.

Another piece of evidence in favor of microlocal coefficient classes is the following trace theorem due to Bao (1989): it concerns the case $c \equiv 1$.

> *Suppose $s > 3 + n/2$ and*
>
> *(i)* $\left[\frac{\partial^2}{\partial t^2} - \Delta - \nabla\sigma \cdot \nabla\right] u = f$ *in* $\mathbb{R}^{n+1}$
>
> *(ii)* $u, f \equiv 0$ *for* $t << 0$
>
> *(iii)* Γ *is a closed conic set containing* $\{(x,\xi) : |\xi_n| \le \epsilon|\xi'|,\ |x| \le \epsilon\}$ *for small* $\epsilon > 0$ *(here* $\xi' = (\xi_1, \ldots \xi_{n-1})$*)*
>
> *(iv)* $u, \nabla\sigma,$ *and* $f \in H^{s-1} \cap H^s_{m\ell}(\Gamma)$
>
> *Then for* $\phi \in C_0^\infty(\mathbb{R}^{n+1})$,
>
> $$\phi u\big|_{x_n=0} \in H^s\ .$$

Thus traces are as regular as the solutions, and particularly as regular as the coefficients allow — but now at a higher degree of smoothness than in the layered case, and only with microlocal restrictions against tangential oscillations.

A great deal remains to be said along these lines, but at least one can imagine that the problems are in some sense "technical." The very delicate dependence of the solution on the principal part of the wave operator — physically on the velocities — poses a quite different sort of difficulty, which forms the second topic of this section. For layered media and plane-wave solutions, the travel-time transformation was available to regularize this dependence — by making the coefficients of the principal part constant! No such possibility exists for $n > 1$. To see how the essential role of the travel-time transformation may nonetheless be filled, we examine once more the one-dimensional case.

Recall from Section 3 that we could write

$$M[r](t) = \int dk\, a(t,k) e^{i\zeta[r](t)k} \hat{r}(k)$$

where $r = dv/dz$ and is presumed to have compact support, and $\zeta[r]$ is the corresponding inverse travel-time function.

Consider more generally a family of oscillatory integral operators of the form

$$G[u]\phi(x) = \int_{\mathbb{R}^n} dy \int_{\mathbb{R}^N} d\theta\, a(u;x,y,\theta) e^{i\phi(u;x,y,\theta)} \phi(y)$$

where the symbol a and phase ϕ are Gateaux-differentiable functions of $u \in C_0^\infty(\mathbb{R})$, to begin with, and assume that G is elliptic. From the calculus of such operators, it follows that the formal perturbation

$$\begin{aligned} \delta G[u;\delta u]\phi(x) &= \int dy \int d\theta \,(\delta a(u,\delta u;x,\theta) \\ &\quad + ia(u;x,\theta)\delta\phi(u,\delta u;x,y,\theta))\, e^{i\phi(u;x,y,\theta)} \phi(y) \end{aligned}$$

is of order higher than that of G, and can be expressed as

$$\delta G[u,\delta u] = G[u]P[u,\delta u] = \tilde{P}[u,\delta u]G[u]$$

where $P, \tilde{P}$ are elliptic pseudo differential operators of order > 0 with symbols linearly dependent on δu. The travel-time transform actually has the form

$$u \mapsto G[u]u\,.$$

Such operators can be said to "have" a canonical relation, even though they are nonlinear — namely, that of $G[u]$. When u changes the canonical relation changes, shifting (infinitesimally: differentiating) the singularities of the image. In order that δG may have the same sort of bounds as G — in terms of δu and of u — it is necessary to compose G with another (necessarily equally irregular) transform, having the inverse canonical relation. That is, we need to find H and Q so that $Q = G \circ H$ has a derivative δQ with the same qualitative estimates as Q itself. It is natural to look for H also in the form of an oscillatory integral, and to expect Q to be pseudo-differential — this is exactly what happens in the case of travel-time. That is, we seek $H[\tilde{u}]$ so that

$$Q[\tilde{u}] = G[H[\tilde{u}]\tilde{u}]H[\tilde{u}]$$

is pseudodifferential, with symbol depending smoothly on $\tilde{u}$.

In the case at hand, $H[\tilde{u}]$ is essentially uniquely specified by this requirement, since its canonical relation is specified. To construct H explicitly, it is necessary to explain first the canonical relation associated to the seismogram operator. Fortunately, the *linearized* seismogram operator has the same canonical relation, and this latter has been discussed extensively — see e.g. Beylkin (1985), Rakesh (1988), for example. The matter is clearest when the oscillatory integral G is expressed as a generalized Radon transform (the viewpoint emphasized by Beylkin).

For simplicity we discuss only the two-dimensional case. Let $R = \{(x_1, x_2) : -X \leq x_1 \leq X, 0 \leq x_2 \leq Z\}$ and suppose that $\mathcal{C}$ is a class of C^1 metrics of the form $c^{-2} \sum_i dx_i dx_i$ in R for which the geodesics between points in R and points in the surface segment $[-X_R, X_R] \times \{0\}$, $x_R < X$, have no conjugate points and lie entirely in R. This hypothesis holds for $c \equiv$ const., hence for a C^2–neighborhood of the constants. This "no-caustics" hypothesis is necessary if results as simple as those to follow are to hold, as is shown in Percell's thesis (1989).

Denoting by $\tau(x, y)$ the geodesic distance from x to y, the *reflection phase* for $y \in R$, $x_r \epsilon [-X_R, X_R]$:

$$\phi(c; x_r; y) = \tau(0, y) + \tau(y, (x_r, 0))$$

is well-defined for $c \in \mathcal{C}$. For $f \in C^\infty(R)$ define

$$G[c]f(x_r, t) = \int_{y:\phi(c;x_r,y)=t} f\ .$$

That is, $G[c]f$ is the generalized Radon transform of f over the equal-time curve $\{y : \phi(c; x_r, y) = t\}$ generated by C. For $c =$ const., the equal-time curves are arcs of ellipses with focii at $(0, 0), (x_r, 0)$ and radii proportional to t.

For smooth c, the impulsive seismogram operator

$$S[c] = u\big|_{x_2=0}$$

where

$$\frac{1}{c^2}\frac{\partial^2 u}{\partial t^2} - \Delta u = \delta(t)\delta(x) \qquad u \equiv 0, \;\; t < 0$$

can be expressed as

$$S[c] = \tilde{p}[c]G[c]c$$

for a suitable pseudodifferential $\tilde{p}[c]$, so S "has the same canonical relation as G."

The class of operators with the inverse canonical relation to that of $G[c]$ all differ from the weighted adjoint operator

$$G^*[c]\tilde{v}(y) = \int_{-X_R}^{X_R} dx_r b(x_r, X_R, y)\tilde{v}(x_r, \phi(c; x_r, y))$$

by composition with the pseudodifferential operator.

Note that $G^*[c]$ maps functions $\tilde{v}(x_r, t)$ of "receiver" and "time" to functions of y, i.e. space — in exploration seismic parlance, $G^*[c]$ is a *migration operator*. The weight b has been included for generality — changing it will change the result by a pseudodifferential operator.

Since $G^*[c]$ depends on c, it is not exactly right. We need an operator defined in terms of functions $\tilde{c}(x_r, t)$, which will function as the equivalent of the travel-time velocity $\tilde{v}$ of Section 3. We define such an operator in terms of an auxiliary problem, as follows.

The "trick" is to construct directly the reflection phase. Note that

$$\phi(1; x_r, y) = |y| + |y_1 - x_r, y_2)| =: \Phi_0(x_r, y) .$$

We will need to control first derivatives of the reflection phase — but Φ_0 is not C^1 near $y = 0$, $y = (x_r, 0)$. So we define instead for $\delta > 0$, $k = 1, 2, \ldots$

$$\begin{aligned} \mathcal{P}_\delta^k \;=\; & \{\Phi \in C^0([-X_R, X_R] \times R) : \\ & \Phi(x_r, y) = \Phi_0(x_r, y) + \Delta\Phi_1(x_r, y), \\ & \Delta\Phi_1 \in C^k([-X_R, X_R] \times R) \text{ and } \\ & \Phi_1(x_r, y) = 0 \text{ if } y_2 < \delta .\} \end{aligned}$$

It is easy to see that if $c \in \mathcal{C}$ satisfies $c(y) \equiv 1$ for $y_2 < \delta$, then $\phi(c; \cdot, \cdot) \in \mathcal{P}_\delta^1$. We give $\mathcal{P}_\delta^k$ the C^k topology in the obvious way.

Pick $\gamma \in C^\infty(\mathbb{R})$ with $\gamma(y) \equiv 0$ for $y_2 < \delta$, $\gamma(y) \equiv 1$ for $y_2 > 2\delta$. For $\tilde{r} \in C^3([-X_R, X_R] \times \mathbb{R}^+)$, $\Phi \in \mathcal{P}_\delta^k$ define

$$\begin{aligned} \Phi_*(\tilde{r}) &= \int_{-X_R}^{X_R} dx_r \tilde{r}(x_r, \Phi(x_r y)) \\ v[\Phi, \tilde{r}] &= 1 + \gamma \Phi_*[\tilde{r}] \end{aligned}$$

We consider the fixed-point problem

$$T_{\tilde{r}}\Phi := \phi(v[\Phi, \tilde{r}]; \cdot, \cdot) = \Phi .$$

If $\Phi = \Phi[\tilde{r}]$ solves this fixed-point problem, then the operator Φ_* has the same canonical relation as $G^*[v[\Phi, \tilde{r}]]$, hence the inverse canonical relation to $G[v[\Phi, \tilde{r}]]$ or $S[v[\Phi, \tilde{r}]]$. Hence

$$\tilde{r} \mapsto S[v[\Phi[\tilde{r}], \tilde{r}]]$$

is "nonlinear pseudodifferential" and has some hope of extending to a differentiable map with coercive derivative on interesting classes.

Note that if $\|\tilde{r}\|_{C^{k+1}}$ is sufficiently small, and $\Phi = \Phi_0 + \Phi_1$ as above with $\|\Phi_1\|_{C^1} \le 1$, then $v[\phi, \tilde{r}]$ is positive, and even a member of $\mathcal{C}$ with $v[\phi, \tilde{r}](y) = 1$ if $y_2 < \delta$. Thus $T_{\tilde{r}}$ maps $\mathcal{P}_\delta^k$ into itself for $\|\tilde{r}\|_{C^{k+1}}$ small enough. Also, $\Phi_*[0] = 0$, so $v[\phi, 0] \equiv 1$ for any $\Phi \in \mathcal{P}_\delta$. Thus $T_0\Phi = \Phi_0$, and therefore T_0 has a unique fixed point, namely Φ_0. If $\Phi, \Phi' \in \mathcal{P}_\delta^k$

$$\begin{aligned} \Phi = \Phi_0 + \Phi_1 , \quad & \Phi' = \Phi_0 + \Phi_1' , \\ \|\Phi_1\|_{C_1} \le 1 , \quad & \|\Phi_1'\|_{C^1} \le 1 , \end{aligned}$$

then it is easy to estimate

$$\|\Phi_*[\tilde{r}] - \Phi_*'[\tilde{r}]\|_{C^k(R)} \le \|\Phi - \Phi'\|_{C^k} \|\tilde{r}\|_{C^{k+1}}$$

so

$$\|v[\Phi,\tilde{r}] - v[\Phi',\tilde{r}]\|_{C^k} \le V\|\tilde{r}\|_{C^{k+1}}\|\Phi_1 - \Phi_1'\|_{C^k}$$

for V depending on δ.

Next we need a few facts about the mapping from a velocity c to its travel-time function $\tau(x,y)$. We will assume that $x_2 = 0$ and $c(y) \equiv 1$ for $y_2 < \delta$. Then for $|x-y| < \delta$, $\tau(x,y) = |x-y|$. For y with $y_2 > \delta$, if $c \in \mathcal{C}$ then $\tau(x,y)$ is a C^1 solution of the eikonal equation

$$c^2|\nabla\tau|^2 = 1\,.$$

If moreover

$$c \in \mathcal{C}_M^0 = \left\{c \in \mathcal{C} : \|\log c\|_{c^0(R)} \le M\right\}$$

then we obtain the obvious estimate

$$\|\tau\|_{C^1(R)} \le C_M^0$$

for a suitable constant $C_M^0 > 0$.

We also need bounds on the second derivatives of τ, which can be got as follows. The Hamiltonian system associated to the eikonal equations is

$$\dot{y} = c^2\eta \quad \dot{\eta} = -c\nabla c|\eta|^2 = -\frac{\nabla c}{c}\,.$$

Denote by $(y(t,y_1),\xi(t,y_1))$ the solution starting (at $t=0$) at

$$(y_1,\delta,\frac{x_1-y_1}{|x-y|},\frac{-y_2}{|x-y|})\,.$$

Then

$$\begin{aligned}\tau(y(t,y_1),x) &= \tau((y_1,\delta),x)+t\\ \nabla\tau(y(t,y_1),x) &= \eta(t,y_1)\,.\end{aligned}$$

From differentiation of the eikonal equation

$$\nabla\tau\cdot\nabla\frac{\partial\tau}{\partial x_i} = -2c^{-3}\frac{\partial c}{\partial x_i}$$

we see that the derivative of $\frac{\partial\tau}{\partial x_i}$ *along the rays* is bounded by $\|c\|_1$. On the other hand,

$$\begin{aligned}\left(\frac{\partial y}{\partial y_1}\right)^{\cdot}(t,y_1) &= \frac{\partial}{\partial y_1}(c^2(y(t,y_1))\eta(t,y_1))\\ &= 2c(y(t,y_1))\left[\nabla c(y(t,y_1))\cdot\frac{\partial y}{\partial y_1}(t,y_1)\right]\cdot\eta(t,y_1)\\ &\quad + c^2(y(t,y_1))\frac{\partial\eta}{\partial y_1}(t,y_1)\\ \left(\frac{\partial\eta}{\partial y_1}\right)^{\cdot}(t,y_1) &= \frac{\partial}{\partial y_1}(\nabla\log c(y(t,y_1)))\\ &= \nabla\nabla\log c(y(t,y_1))\cdot\frac{\partial y}{\partial y_1}(t,y_1)\end{aligned}$$

whence a standard O.D.E. estimate gives

$$\left|\frac{\partial y}{\partial y_1}(t,y_1)\right| \le \text{const.}\ \|\log c\|_{C^2(R)}$$

when $y(t,y_1) \in R$. The assumption that no conjugate points exist implies that $(t,y_1) \mapsto y(t,y_1)$ is a global coordinate system in R, so $\frac{\partial y}{\partial y_1}$ and $\dot{y}$ are linearly independent. Moreover, it is possible to show that the condition number of

$$\begin{pmatrix} \frac{\partial y}{\partial y_1} \\ \dot{y} \end{pmatrix}$$

is

$$\le C(1 - t\|\nabla \log c\|_{C^1})^{-1}$$

(this is an explicit bound guaranteeing global geodesic coordinates). Since

$$\left(\frac{\partial \eta}{\partial y_1}\right)^{\cdot} = \left(\nabla\nabla\tau \cdot \frac{\partial y}{\partial y_1}\right)^{\cdot} = \nabla\nabla \log c \cdot \frac{\partial y}{\partial y_1}$$

the condition number estimate shows that

$$\|\nabla\nabla\tau\|_{C^0} \le C\left[(1 - t\|\nabla \log c\|_{C^1})^{-1}\|\nabla \log c\|_{C^1} + 1\right] .$$

Define

$$\mathcal{C}^2_{M,P} = \left\{c \in \mathcal{C}^0_M : \|\nabla \log c\|_{C^1} \le P\right\} .$$

Then if P is small enough, there is a constant $K_2 > 0$ so that

$$c \in \mathcal{C}^2_{M,P} \Rightarrow \|\phi(c;\cdot,\cdot) - \Phi_0\|_{C^2} \le K_2 .$$

Next, suppose $c_2, c_2 \in \mathcal{C}^2_{M,P}$. Then

$$\nabla\tau_1 \cdot \nabla\frac{\partial \tau_1}{\partial x_i} - \nabla\tau_2 \cdot \nabla\frac{\partial \tau_2}{\partial x_i} = 2\left(c_2^{-3}\frac{\partial c_2}{\partial x_i} - c_1^{-3}\frac{\partial c_1}{\partial x_i}\right)$$

$$= \nabla\tau_1 \cdot \nabla\left(\frac{\partial \tau_1}{\partial x_i} - \frac{\partial \tau_2}{\partial x_i}\right) + (\nabla\tau_1 - \nabla\tau_2)\cdot\nabla\frac{\partial \tau_2}{\partial x_i} .$$

Integrating along the rays for τ_1 and using $\tau_1 = \tau_2$ at $y_2 = \delta$ we get

$$\|\tau_1 - \tau_2\|_{C^1} \le \frac{1}{2}L\|c_1 - c_2\|_{C^1}$$

where L is a function of M and P (and in particular of $\|\nabla \log c_i\|_{C^1}$), whence

$$\|\phi(c_1;\cdot,\cdot) - \phi(c_2;\cdot,\cdot)\|_{C^1} \le L\|c_1 - c_2\|_{C^1} .$$

Now we are ready to string the various facts together. Set

$$\mathcal{B} = \mathcal{P}^2_\delta \cap \{\|\Phi - \Phi_0\|_{C^2} \le K_2\} .$$

For $\Phi \in \mathcal{B}$,

$$\begin{aligned}\|v[\Phi,\tilde{r}]-1\|_{C^2} &\le \|\tilde{v}[\Phi,\tilde{r}]-\tilde{v}[\Phi_0,\tilde{r}]\|_{C^2}+\|\tilde{v}[\Phi_0,\tilde{r}]-\tilde{v}[\Phi_0,0]\|_{C^2}\\ &\le VK_2\|r\|_{C^3}+\|\Phi_*[\tilde{r}]\|_{C^2}\\ &\le (VK_2+W)\|r\|_{C^3}\end{aligned}$$

where W is an upper bound for the length of the curves $x_r \mapsto (x_r, \Phi(y, x_r))$ as y ranges over R and Φ over $\mathcal{B}$.

Consequently there exists $Q > 0$ so that if $\|\tilde{r}\|_{C^3} \le Q$, then $\Phi \in \mathcal{B}$ implies

$$v[\Phi,\tilde{r}] \in \mathcal{C}^2_{M,P}\,.$$

Q depends on V, M, and P (and on K_2 and W, though these depend on M and P.) Consequently, if $\Phi \in \mathcal{B}$, $\|r\|_3 \le Q$, then

$$\|\phi(v[\Phi,\tilde{r}];\cdot,\cdot)-\Phi_0\|_{C^2} \le K_2\,,$$

i.e. $T_{\tilde{r}}$ maps $\mathcal{B}$ into itself. Next note that for $\Phi_1, \Phi_2 \in \mathcal{B}$,

$$\begin{aligned}\|T_{\tilde{r}}[\Phi_1]-T_{\tilde{r}}[\Phi_2]\|_{C^1} &\le L\|v[\Phi_1,\tilde{r}]-v[\Phi_2,\tilde{r}]\|_{C^1}\\ &\le VL\|\tilde{r}\|_{C^2}\|\Phi_1-\Phi_2\|_{C^1}\,.\end{aligned}$$

Thus by making Q possibly smaller yet, we can arrange that $T_{\tilde{r}}$ is a contraction on $\mathcal{B}$ — but only in the sense of C^1, not C^2. So we don't get a solution of the fixed point problem in $\mathcal{B}$ — nonetheless, the sequence of iterates $\{T_{\tilde{r}}^n[\Phi_r]\}$ is

(a) bounded in C^2

(b) convergent in C^1 to a C^1 solution of $T_{\tilde{r}}\Phi = \Phi$.

This result is only a glimpse of the actual state of affairs concerning the "multi-dimensional travel time transformation," which the author expects to occupy a central position in the study of wave propagational inverse problems.

Acknowledgements. I am pleased to thank the organizers of the Leesburg Workshop on Random Media, Robert Kohn and Graeme Milton, for the opportunity to present this and related material at the Workshop, and to thank SIAM for its support. The work reported here was partially supported by the National Science Foundation under grant MCS 86-03614, and by the Office of Naval Research under contracts N00014-K-85-0725 and N00014-J-89-1115. The hospitality of Professor Guy Chavent and the program CEREMADE, Université de Paris–Dauphine, during the writing, is gratefully acknowledged.

REFERENCES

1. A. BAMBERGER, G. CHAVENT and P. LAILLY, *About the stability of the inverse problem in 1-D wave equation—application to the interpretation of seismic profiles*, Appl. Math. Opt., 5 (1979), pp. 1-47.

2. G. BAO, work in progress, preprint (1989).

3. M. BEALS and M. REED, *Propagation of singularities for hyperbolic pseudodifferential operators with nonsmooth coefficients*, Comm. Pure. Appl. Math., 35 (1982), pp. 169-184.

4. M. BEALS and M. REED, *Microlocal regularity theorems for nonsmooth pseudodifferential operators and applications to nonlinear problems*, Trans. Am. Math. Soc., 285 (1984), pp. 159-184.

5. G. BEYLKIN, *Imaging of discontinuities in the inverse scattering problem by inversion of a causal generalized Radon transform*, J. Math. Phys., 26 (1985), pp. 99-108.

6. R. BURRIDGE, G. PAPANICOLAU, P. SHENG, and B. WHITE, *Probing a random medium with a pulse*, SIAM J. Appl. Math., 49 (2) (1989).

7. J. CARAZZONE and W. SYMES, *Velocity inversion by coherency optimization*, preprint (1989).

8. R. COURANT and D. HILBERT, *Methods of Mathematical Physics*, J. Wiley/Interscience, New York, 1962.

9. R.M. LEWIS, *An inverse problem for determination of velocity and source in an acoustic medium*, Ph.D. thesis, Department of Mathematical Sciences, Rice University, Houston, Texas, 1989.

10. J. MORLET, G. ARENS, E. FOURGEAU and D. GIARD, *Wave propagation and sampling theory, Parts I and II*, Geophysics, 47 (1982), pp. 203-236.

11. C. PERCELL, *The effect of caustics in acoustic inverse scattering experiments*, Ph.D. thesis, Department of Mathematical Sciences, Rice University, Houston, Texas, 1989.

12. RAKESH, *A linearized inverse problem for the wave equation*, Comm. P.D.E., 13 (1988), pp. 573-601.

13. J. RAUCH, *Singularities of solutions to semilinear wave equations*, J. Math. Pure Appl., 58 (1979), pp. 299-308.

14. P. SACKS and W. SYMES, *Uniqueness and continuous dependence for a multidimensional hyperbolic inverse problem*, Comm. P.D.E., 10 (1985), pp. 635-676.

15. Z. SUN, *Uniqueness theorems for inverse problems in wave propagation and electrostatics*, Ph.D. thesis, Department of Mathematics, University of California, Los Angeles, 1987.

16. T. SUZUKI, *Stability of an inverse hyperbolic problem*, Inverse Problems, 4 (1988), pp. 273-299.

17. W. SYMES, *On the relation between coefficient and boundary values for solutions of Webster's horn equation*, SIAM J. Appl. Math, 17 (1986a), pp. 1400-1420.

18. W. SYMES, *Stability properties for the velocity inversion problem*, Proc. SEG/ SIAM/SPE Conference on Seismic Exploration and Reservoir Modeling, ed. W. Fitzgibbon, SIAM, Philadelphia (1986b).

19. W. SYMES, *Layered velocity inversion: A model problem from reflection seismology*, Technical Report 88-13, Department of Mathematical Sciences, Rice University, Houston, Texas (1988).

CHAPTER 6

H-measures and Small Amplitude Homogenization

Luc Tartar*

Introduction

Around 1974, after having obtained with F. MURAT some mathematical results which extended some earlier ones of S. SPAGNOLO [**9**,**10**] on a subject that was not yet called homogenization, I found in a course of theoretical physics a paragraph on the conductivity of mixtures which was quite puzzling [**6**]. The mathematical results that we knew showed that, except in dimension 1, one could not deduce the conductivity of a mixture from the proportions and conductivities of the constituents; what could then be the meaning of the formulas derived in that physics course?

Of course our proofs were a little more technical than the ones we taught a few years after [**8**,**11**] and I had not thought yet of the method for obtaining bounds [**12**] that we successfully applied for mixtures of two isotropic materials [**13**], but it was clear that the tacit assumption that the resulting mixture was an isotropic material could not imply that its conductivity could be computed. The conductivity could lie in an interval whose precise bounds were not known but which was certainly not reduced to a point.

More puzzling than the fact of solving a problem which had no definite answer was the method used for deriving the formula. First a special problem was solved explicitly, that of a sphere imbedded in an infinite medium, all materials being isotropic. Then some formulas valid for this special case were manipulated in such a way as to finally guess a formula that could be meaningful in all cases. Did it mean that the formula should be applied only to spherical particles that were far apart? It was not

* Department of Mathematics, Carnegie Mellon University, Pittsburgh, PA 15213-3890.

mentioned and one could not either find a reference to any experimental data showing how good or bad that formula was. If it had been the case that the derived formula was experimentally accurate, what could a mathematician learn from an argument which had no logical foundation even if it did give the right answer in a special case?

A year later, I learned from I. BABUSKA both the term homogenization itself and the fact that there were plenty of similarly strange formulas in the physics litterature. However his examples were of the opposite kind because he was considering a definite periodic micro-geometry and there was one mathematical answer in that case, while the physics litterature contained many different ones [**3**].

At the end of 1986, I decided to open again that physics book in order to compare the formulas it contained with the optimal bounds that were known in some very restrictive cases. Having learned more about the way physicists think I was then able to go beyong the irrational arguments and it was only then that I realized that L.D. LANDAU & E.M. LIFSCHITZ [**6**] were deriving a simple formula that was supposed to hold in the case of small variations of the conductivity. In the case of two-component mixtures this argument was actually consistent with the optimal bounds first introduced by Z. HASHIN & S. SHTRIKMAN, and I then checked with G. FRANCFORT and F. MURAT that one could deduce their general claim by using some technical lemmas that I had derived with F. MURAT for proving more general bounds [**13**].

Of course what we had then proved was a precise mathematical statement which was our interpretation of what L.D. LANDAU & E.M. LIFSCHITZ had claimed in [**6**]. It was not based on any explicit computation corresponding to a special geometry, but relied on technical arguments of compensated compactness that we did not know how to generalize to other situations, with the exception of the extension to linear elasticity performed by G. FRANCFORT & F. MURAT [**4**]. Based on a preceding hint on quadratic corrections in homogenization [**14**], I developped in the following months a new mathematical tool adapted to that precise question of small amplitude homogenization, but for a more general setting. It is a natural extension for general sequences of an argument which I have first heard from G. MILTON in the case of periodic micro-geometries and it can be connected to the work of many specialists, among them M. AVELLANEDA, R.V. KOHN and G. MILTON [**1**,**2**,**5**,**7**]. In the beginning of 1987 I had mentioned these results to J.B. KELLER who confirmed my impression that it was similar to results obtained under Born's approximation. It was only much later that I obtained other applications like transport properties of oscillations and concentration effects that should be useful for other problems in homogenization like evolution of mixtures.

As I have described in [**15**] some general mathematical properties of what I have called H-measures and the particular example of small amplitude homogenization for a diffusion equation, I want to describe here the results and what would be the extension to the isotropic case in linear elasticity. I refer to [**15**] for the way to transform into theorems the following computations by using the definition of H-measures together with the localization principle.

1. The diffusion equation

One finds in [**6**] a formula for the effective conductivity a_{eff} of a mixture of isotropic diffusive materials with conductivity a(x) which is computed in the following way. One denotes by $\mathbf{a}_0$ the local average of a(x) and by σ^2 the average of $(a-\mathbf{a}_0)^2$ and then the formula can be written as

$$a_{eff} = \mathbf{a}_0 - \frac{\sigma^2}{3\mathbf{a}_0}. \tag{1.1}$$

This formula is derived in dimension 3 and is supposed to apply to the case where the oscillations of a(x) and thus σ are small.

The following theorem, consequence of a more general result proved in [**15**], can be considered as the proof that formula (1.1) is indeed a good guess.

THEOREM 1.1. *Let Ω be a bounded open set of R^N, let $\mathbf{a}_0$ be a positive continuous function in Ω and let b_ε and c_ε be sequences converging respectively to b_0 and c_0 in $L^\infty(\Omega)$ weak * with $(b_\varepsilon - b_0)^2$ converging to σ^2 in $L^\infty(\Omega)$ weak *.*

Assume that we have extracted a subsequence such that for small values of the parameter γ the sequence of conductivity tensors

$$A_\varepsilon = (\mathbf{a}_0 + \gamma b_\varepsilon + \gamma^2 c_\varepsilon)I \tag{1.2}$$

converges in the sense of homogenization to an effective conductivity tensor A_{eff} (which is analytic in γ).

Then one has the following formula

$$A_{eff} = \mathbf{a}_0 I + \gamma b_0 I + \gamma^2\left(c_0 I - \frac{1}{\mathbf{a}_0}\mathbf{M}\right) + O(\gamma^3) \tag{1.3}$$

where $\mathbf{M}$ is a symmetric nonnegative matrix satisfying

$$\mathrm{Trace}(\mathbf{M}) = \sigma^2. \tag{1.4}$$

Remark 1.2. If the effective medium is isotropic for all value of γ then $\mathbf{M}$ is proportional to the identity matrix so that in that case one as

$$\mathbf{M} = \frac{\sigma^2}{N} I \tag{1.5}$$

which is indeed a way to interpret what formula (1.1) claims.

Remark 1.3. As pointed out in [**15**] the assumption of continuity on $\mathbf{a}_0$ can be dropped and the theorem extends to general conductivity tensors. The method of H-measures constructs a nonnegative measure μ on $\Omega \mathrm{x} S^{N-1}$ associated to a subsequence of $(b_\varepsilon - b_0)$ and then expresses the correction $\mathbf{M}/\mathbf{a}_0$ as an integral of μ on the sphere S^{N-1}.

Remark 1.4. The proof actually gives a characterization of the sequences b_ε which result in an isotropic correction $\mathbf{M}$, namely those for which the H-measure μ corresponding to a subsequence of $b_\varepsilon - b_0$ satisfies

$$<\mu.\phi\otimes\xi_i\xi_j> = \delta_{ij}\int_\Omega \phi(x)\kappa(x)dx \text{ for all i,j and } \phi \in C_c(\Omega) \qquad (1.6)$$

for some function κ, which is then σ^2/N. This may result from symmetries of the micro-geometry which are obviously seen by the H-measure μ.

2. The linear elasticity system

We consider now a mixture of isotropic linear elastic materials, i.e. where the stress tensor σ has the form

$$\sigma = 2\mu e + \lambda(\mathrm{Tr} e)I \qquad (2.1)$$

where Tr denotes the trace and where the linearized strain tensor $\mathbf{e}$ is computed from the displacement $\mathbf{u}$ by

$$\mathbf{e}_{ij} = \frac{1}{2}\left(\frac{\partial \mathbf{u}_i}{\partial x_j} + \frac{\partial \mathbf{u}_j}{\partial x_i}\right). \qquad (2.2)$$

For sequences λ_ε and μ_ε satisfying some natural constraints, one can extract a subsequence which converges in the sense of homogenization to an effective medium which may not be isotropic and whose constitutive equation may have the general form

$$\sigma_{ij} = \sum_{kl} C_{ijkl} e_{kl}. \qquad (2.3)$$

We want to consider sequences of the form

$$\lambda_\varepsilon = \lambda + \gamma\lambda_1^\varepsilon + \gamma^2\lambda_2^\varepsilon\ ;\ \mu_\varepsilon = \mu + \gamma\mu_1^\varepsilon + \gamma^2\mu_2^\varepsilon \qquad (2.4)$$

where λ and μ are positive continuous functions and

$$\lambda_j^\varepsilon \to \lambda_j \text{ and } \mu_j^\varepsilon \to \mu_j \text{ in } L^\infty(\Omega) \text{ weak * for } j = 1,2 \qquad (2.5)$$

and we are interested in the expansion with respect to γ of the effective elasticity coefficients C_{ijkl}. The coefficient of γ^2 will contain a correction obtained by integrating some H-measure on the unit sphere S^{N-1}, but as we will consider subsequences of the vector valued function $(\mu^\varepsilon_1-\mu_1,\lambda^\varepsilon_1-\lambda_1)$ the H-measure ν will be a nonnegative hermitian matrix whose entries ν^{ij} are measures on $\Omega\times S^{N-1}$.

The procedure starts in a classical way, using the analyticity in the parameter γ. We assume Dirichlet conditions, choose a function **u** belonging to $(H^1_0(\Omega))^N$ and relate **e** to **u** by (2.2); we construct a sequence e^ε of the form

$$e^\varepsilon = \mathbf{e} + \gamma e_1^\varepsilon + \gamma^2 e_2^\varepsilon + \ldots \tag{2.6}$$

where e^ε_1 and e^ε_2 converge weakly to 0, have the form (2.2), and satisfy

$$\mathrm{Div}[2\mu e_1 + \lambda(\mathrm{Tr}e_1)I + 2\mu_1^\varepsilon \mathbf{e} + \lambda_1^\varepsilon(\mathrm{Tr}\mathbf{e})I] = \mathrm{Div}[f] \tag{2.7}$$

$$\mathrm{Div}[2\mu e_2 + \lambda(\mathrm{Tr}e_2)I + 2\mu_1^\varepsilon e_1^\varepsilon + \lambda_1^\varepsilon(\mathrm{Tr}e_1^\varepsilon)I + 2\mu_2^\varepsilon \mathbf{e} + \lambda_2^\varepsilon(\mathrm{Tr}\mathbf{e})I] = \mathrm{Div}[g] \tag{2.8}$$

where f and g are compatible with the fact that the sequences e^ε_1 and e^ε_2 converge weakly to 0, i.e. $f = 2\mu_1\mathbf{e} + \lambda_1(\mathrm{Tr}\mathbf{e})I$ and $g = 2\mu_2\mathbf{e} + \lambda_2(\mathrm{Tr}\mathbf{e})I + \lim_\varepsilon\{2\mu^\varepsilon_1 e^\varepsilon_1 + \lambda^\varepsilon_1(\mathrm{Tr}e^\varepsilon_1)I\}$; as usual $w = \mathrm{Div}[\tau]$ means $w_i = \Sigma_j \partial\tau_{ij}/\partial x_j$ for each i. In order to compute the coefficient of γ^2 in the expansion of C_{ijkl} we need to express the limit of $2\mu^\varepsilon_1 e^\varepsilon_1 + \lambda^\varepsilon_1(\mathrm{Tr}e^\varepsilon_1)I$ in terms of **e**.

This is where the method described in [**15**] comes into play. As this quantity is quadratic its limit can be computed using weak limits and a H-measure corresponding to a subsequence of $(\mu^\varepsilon_1-\mu_1,\lambda^\varepsilon_1-\lambda_1,e^\varepsilon_1)$. Using (2.2) and (2.7) and the localization principle, this H-measure can in turn be expressed in terms of a H-measure corresponding to a subsequence of $(\mu^\varepsilon_1-\mu_1,\lambda^\varepsilon_1-\lambda_1)$. Some algebraic computations have to be performed and the first step consists, for each vector ξ of the unit sphere S^{N-1}, in solving the system

$$[2\mu e_1 + \lambda(\mathrm{Tr}e_1)I + 2\mu_1\mathbf{e} + \lambda_1(\mathrm{Tr}\mathbf{e})I]\xi = 0 \tag{2.9}$$

with e_1 having the form

$$e_1 = \frac{1}{2}(a\otimes\xi + \xi\otimes a). \tag{2.10}$$

For each ξ this will give e_1 depending linearly in λ_1 and μ_1 and the second step consists in using this dependence to express $2\mu_1 e_1 + \lambda_1(\mathrm{Tr}e_1)I$ as a quadratic expression in λ_1 and μ_1. The third step will give the required limit by integrating on the sphere S^{N-1} a similar expression involving the H-measure ν.

In order to solve explicitly (2.9)-(2.10), let us denote

$$X_1 = 2\mu_1 e + \lambda_1(\mathrm{Tr}e)I. \tag{2.11}$$

As $(e_1\xi.\xi) = (a.\xi) = \mathrm{Tr}e_1$ by (2.10), (2.9) implies

$$\mathrm{Tr}e_1 = -\frac{(X_1\xi.\xi)}{2\mu+\lambda}, \tag{2.12}$$

$$a = -\frac{1}{\mu}X_1\xi + \frac{\mu+\lambda}{\mu(2\mu+\lambda)}(X_1\xi.\xi)\xi \tag{2.13}$$

and

$$e_1 = -\frac{1}{2\mu}[X_1\xi\otimes\xi + \xi\otimes X_1\xi] + \frac{\mu+\lambda}{\mu(2\mu+\lambda)}(X_1\xi.\xi)\xi\otimes\xi, \tag{2.14}$$

or

$$e_1 = \mu_1\left\{-\frac{e\xi\otimes\xi+\xi\otimes e\xi}{\mu} + 2\frac{(\mu+\lambda)(e\xi.\xi)}{\mu(2\mu+\lambda)}\xi\otimes\xi\right\} + \lambda_1\left\{-\frac{(\mathrm{Tr}e)}{2\mu+\lambda}\xi\otimes\xi\right\}. \tag{2.15}$$

This gives

$$\begin{cases} 2\mu_1 e_1 + \lambda_1(\mathrm{Tr}e_1)I = 2\mu_1^2\left\{-\frac{e\xi\otimes\xi+\xi\otimes e\xi}{\mu} + 2\frac{(\mu+\lambda)(e\xi.\xi)}{\mu(2\mu+\lambda)}\xi\otimes\xi\right\} + \\ \qquad 2\mu_1\lambda_1\left\{-\frac{(\mathrm{Tr}e)}{2\mu+\lambda}\xi\otimes\xi - \frac{(e\xi.\xi)}{2\mu+\lambda}I\right\} + \\ \qquad \lambda_1^2\left\{-\frac{(\mathrm{Tr}e)}{2\mu+\lambda}I\right\} \end{cases} \tag{2.16}$$

which implies

$$\mathrm{Tr}[2\mu_1 e_1 + \lambda_1(\mathrm{Tr}e_1)I] = -\frac{2\mu_1+N\lambda_1}{2\mu+\lambda}\{2\mu_1(e\xi.\xi) + \lambda_1(\mathrm{Tr}e)\}. \tag{2.17}$$

The expansion of C_{ijkl} in powers of γ has the form

$$C_{ijkl} = 2(\mu+\gamma\mu_1+\gamma^2\mu_2)\delta_{ik}\delta_{jl} + (\lambda+\gamma\lambda_1+\gamma^2\lambda_2)\delta_{ij}\delta_{kl} + \gamma^2 D_{ijkl} + O(\gamma^3) \tag{2.18}$$

where each term in (2.16) contributes to the coefficient D_{ijkl} and this will involve the H-measure ν associated to a subsequence of $(\mu^\varepsilon{}_1-\mu_1,\lambda^\varepsilon{}_1-\lambda_1)$. In order to express these contributions we will use the following notation

NOTATION 2.1. Let a_{ij}, b_{ij} and c_{ij} be the integral of $\xi_i\xi_j$ on the sphere S^{N-1} with respect to the measures ν^{11}, $\mathrm{Re}\nu^{12}$ and ν^{22}. Let a_{ijkl} be the integral of $\xi_i\xi_j\xi_k\xi_l$ on the sphere S^{N-1} with respect to the measure ν^{11}.

The contribution to D_{ijkl} corresponding to the coefficient of $\mu_1{}^2$ in (2.16) is then

$$-\frac{1}{\mu}(\delta_{ik}a_{jl} + \delta_{il}a_{jk} + \delta_{jk}a_{il} + \delta_{jl}a_{ik}) + \frac{4(\mu+\lambda)}{\mu(2\mu+\lambda)}a_{ijkl}, \tag{2.19}$$

the contribution corresponding to the coefficient of $\mu_1\lambda_1$ is

$$-\frac{2}{2\mu+\lambda}\delta_{kl}b_{ij} - \frac{2}{2\mu+\lambda}\delta_{ij}b_{kl}, \tag{2.20}$$

and the contribution corresponding to the coefficient of $\lambda_1{}^2$ is

$$-\frac{1}{2\mu+\lambda}\delta_{ij}\delta_{kl}(\mathrm{Tr}c). \tag{2.21}$$

The correction D_{ijkl} being the sum of the expressions (2.19), (2.20) and (2.21) we consider only the special cases corresponding to $l = k$ or j (without summing yet on repeated indices)

$$D_{ijkk} = -\frac{2}{\mu}(\delta_{ik}a_{jk} + \delta_{jk}a_{ik}) + \frac{4(\mu+\lambda)}{\mu(2\mu+\lambda)}a_{ijkk} - \frac{2}{2\mu+\lambda}b_{ij} - \frac{2}{2\mu+\lambda}\delta_{ij}b_{kk} - \frac{1}{2\mu+\lambda}(\mathrm{Tr}c)$$

and

$$D_{ijkj} = -\frac{1}{\mu}(\delta_{ik}a_{jj} + \delta_{ij}a_{jk} + \delta_{jk}a_{ij} + a_{ik}) + \frac{4(\mu+\lambda)}{\mu(2\mu+\lambda)}a_{ijkj} - \frac{2}{2\mu+\lambda}\delta_{kj}b_{ij} - \frac{2}{2\mu+\lambda}\delta_{ij}b_{kj} - \frac{1}{2\mu+\lambda}\delta_{ij}\delta_{kj}(\mathrm{Tr}c).$$

Summing now on repeated indices we obtain

$$\sum_{k=1}^{N} D_{ijkk} = -\frac{4}{2\mu+\lambda}a_{ij} - \frac{2N}{2\mu+\lambda}b_{ij} - \frac{2(\mathrm{Tr}b)+N(\mathrm{Tr}c)}{2\mu+\lambda}\delta_{ij} \text{ for all } i,j \tag{2.22}$$

and

$$\sum_{j=1}^{N} D_{ijkj} = -\frac{2N\mu+(N-2)\lambda}{\mu(2\mu+\lambda)}a_{ik} - \frac{4}{2\mu+\lambda}b_{ik} - \left(\frac{(\mathrm{Tr}a)}{\mu} + \frac{(\mathrm{Tr}c)}{2\mu+\lambda}\right)\delta_{ik} \text{ for all } i,k. \tag{2.23}$$

The preceding formulas are general but if we assume now that the effective material is isotropic for all values of γ, this corresponds to the case

$$D_{ijkl} = 2M\delta_{ik}\delta_{jl} + \Lambda\delta_{ij}\delta_{kl} \tag{2.24}$$

and we can identify M and Λ. Indeed in this special case the formulas (2.22) and (2.23) become

$$(2M+N\Lambda)\delta_{ij} = -\frac{4}{2\mu+\lambda}a_{ij} - \frac{2N}{2\mu+\lambda}b_{ij} - \frac{2(\mathrm{Tr}b)+N(\mathrm{Tr}c)}{2\mu+\lambda}\delta_{ij} \text{ for all } i,j \tag{2.25}$$

and

$$(2NM + \Lambda)\delta_{ik} = -\frac{2N\mu+(N-2)\lambda}{\mu(2\mu+\lambda)}a_{ik} - \frac{4}{2\mu+\lambda}b_{ik} - \left(\frac{(Tra)}{\mu}+\frac{(Trc)}{2\mu+\lambda}\right)\delta_{ik} \text{ for all i,k.} \quad (2.26)$$

By summing (2.25) for i = j and (2.26) for i = k we obtain

$$N(2M+N\Lambda) = -\frac{4(Tra)+4N(Trb)+N^2(Trc)}{2\mu+\lambda} \quad (2.27)$$

and

$$N(2NM + \Lambda) = -\frac{4N\mu+2(N-1)\lambda}{\mu(2\mu+\lambda)}(Tra) - \frac{4}{2\mu+\lambda}(Trb) - \frac{1}{2\mu+\lambda}(Trc) \quad (2.28)$$

which give for $2M$ and Λ the values

$$2M = -\frac{4(N+1)\mu+2N\lambda}{N(N+1)\mu(2\mu+\lambda)}(Tra) \quad (2.29)$$

and

$$\Lambda = \frac{2\lambda}{N(N+1)\mu(2\mu+\lambda)}(Tra) - \frac{4}{N(2\mu+\lambda)}(Trb) - \frac{1}{2\mu+\lambda}(Trc). \quad (2.30)$$

Notice that in the case N = 1 (2.27) and (2.28) only give the value of $2M+\Lambda$ which is obviously the only relevant quantity.

In order to state the result that we have obtained in a simpler way we will use (2.27) and (2.29) and observe that the traces of the matrices a, b and c can be expressed as weak limits of quadratic quantities in μ^{ε}_1 and λ^{ε}_1. More precisely $(\mu^{\varepsilon}_1)^2$ converges weakly to $\mu_1{}^2$ + (Tra) and $(2\mu^{\varepsilon}_1+N\lambda^{\varepsilon}_1)^2$ converges weakly to $(2\mu_1+N\lambda_1)^2$ + {4(Tra)+4N(Trb)+N^2(Trc)}. By using the method described in **[15]** we have then proved the following result

THEOREM 2.2. *If a sequence of isotropic elastic mixtures with parameters μ_ε and λ_ε of the form (2.4)-(2.5) converges in homogenization to an isotropic effective medium, then its parameters satisfy*

$$\mu_{eff} = \mu + \gamma\mu_1 + \gamma^2(\mu_2+M) + O(\gamma^3) \text{ and } \lambda_{eff} = \lambda + \gamma\lambda_1 + \gamma^2(\lambda_2+\Lambda) + O(\gamma^3) \quad (2.31)$$

with $2M$ and Λ given by

$$2M = -\frac{4(N+1)\mu+2N\lambda}{N(N+1)\mu(2\mu+\lambda)}\lim_{\varepsilon\to 0}\{(\mu_1^{\varepsilon}-\mu_1)^2\} \quad (2.32)$$

and

$$(2M+N\Lambda) = -\frac{1}{N(2\mu+\lambda)}\lim_{\varepsilon\to 0}\{(2\mu_1^\varepsilon + N\lambda_1^\varepsilon - 2\mu_1 - N\lambda_1)^2\}. \tag{2.33}$$

Remark 2.3. We have also found in formulas (2.25) and (2.26) some simple necessary conditions for obtaining an isotropic correction at second order. For N ≥ 3 they imply that the matrices a and b are proportional to I and this can be expressed as a condition on the H-measure ν as

$$<\nu^{11}.\phi\otimes\xi_i\xi_j> = \delta_{ij}\int_\Omega \phi(x)f(x)dx \text{ for all i,j and } \phi \in C_c(\Omega) \tag{2.34}$$

and

$$<\mathrm{Re}\,\nu^{12}.\phi\otimes\xi_i\xi_j> = \delta_{ij}\int_\Omega \phi(x)g(x)dx \text{ for all i,j and } \phi \in C_c(\Omega) \tag{2.35}$$

for some bounded functions f and g. For N = 2 we can only deduce that

$$<2\nu^{11} + N\mathrm{Re}\,\nu^{12}.\phi\otimes\xi_i\xi_j> = \delta_{ij}\int_\Omega \phi(x)h(x)dx \text{ for all i,j and } \phi \in C_c(\Omega) \tag{2.36}$$

for a bounded function h.

Of course we can also deduce some information on the moments of order 4 of ν^{11} by using directly the formula for D_{ijkl}.

Remark 2.4. These computations may have been already written elsewhere and certainly others can be done in the context of general symmetries for crystals (or quasi-crystals). The only new idea is that there is now a precise mathematical tool described in **[15]** which explains how to transform these computations into general theorems, without having to make any symmetry or periodicity assumptions for the composition.

Acknowledgements
This work is supported in part by NSF grant DMS-8803317.

References

1. M. AVELLANEDA. Iterated homogenization, differential effective medium theory and applications, Comm. Pure Appl. Math., 40 (1987) no 5, 527-554.

2. M. AVELLANEDA. Optimal bounds and microgeometries for elastic two-phase composites, S.I.A.M. J. Appl. Math., 47 (1987) no 6, 1216-1228.

3. I. BABUSKA Homogenization and its application. Mathematical and computational problems. Numerical solution of partial differential equations-III. SYNSPADE 1975 (HUBBARD B. ed), Academic Press, 1976, 89-116.

4. G. FRANCFORT and F. MURAT. Homogenization and optimal bounds in linear elasticity, Arch. Rat. Mech. Anal., 94 (1986) no 4, 307-334.

5. R.V. KOHN and G.W. MILTON. On bounding the effective conductivity of anisotropic composites, Homogenization and effective moduli of materials and media, (J.L. Ericksen, D. Kinderlehrer, R.V. Kohn, J.L. Lions eds), 97-125, The IMA Volumes in Math. and its Applic., 1, Springer-Verlag, 1986.

6. L.D. LANDAU and E.M. LIFSCHITZ. Electrodynamics of continuous media, Pergamon Press, 1984.

7. G.W. MILTON. Modelling the properties of composites by laminates, Homogenization and effective moduli of materials and media, (J.L. Ericksen, D. Kinderlehrer, R.V. Kohn, J.L. Lions eds), 150-174, The IMA Volumes in Math. and its Applic., 1, Springer-Verlag, 1986.

8. F. MURAT. H-convergence, Séminaire d'analyse fonctionnelle et numérique de l'Université d'Alger. 1977-78, 34 p.

9. S. SPAGNOLO. Sul limite delle soluzioni di problemi di Cauchy relativi all' equazione del calore, Ann. Scuola Norm. Sup. Pisa (3) 21 (1967), 657-699.

10. S. SPAGNOLO. Sulla convergenza di soluzioni di equazioni paraboliche ed ellitiche, Ann. Scuola Norm. Sup. Pisa (3) 22 (1968), 571-597.

11. L. TARTAR. Cours Peccot, Collège de France, Paris, 1977. Unpublished, partially written in F. MURAT. H-convergence, 1977-78, Alger.

12. L. TARTAR. Estimations de coefficients homogénéisés, Computing methods in applied sciences and engineering, I, 364-373, Lecture Notes in Math., 704, Springer, 1979.

13. L. TARTAR. Estimations fines de coefficients homogénéisés, Ennio de Giorgi Colloquium, (P. Krée ed.), 168-187, Research Notes in Mathematics 125, Pitman, (1985).

14. L. TARTAR. Remarks on homogenization, Homogenization and effective moduli of materials and media, (J.L. Ericksen, D. Kinderlehrer, R.V. Kohn, J.L. Lions eds), 228-246, The IMA Volumes in Math. and its Applic., 1, Springer-Verlag, 1986.

15. L. TARTAR. H-measures, a new approach for studying homogenization, oscillations and concentration effects in partial differential equations, to appear, Proceedings of the Royal Society of Edinburgh.

CHAPTER 7

The Overall Behaviour of a Nonlinear Matrix Reinforced by Rigid Inclusions or Weakened by Cavities

J. R. Willis*

Abstract. A generalization to nonlinear composites of the Hashin-Shtrikman variational structure is briefly reviewed and then specialized to a nonlinear two-component composite. The nonlinear algebraic equations that the formalism generates are solved explicitly, for general matrix behaviour, for a composite consisting of a matrix containing rigid inclusions. Self-consistent prescriptions are discussed and illustrated by reference to results generated for a particular type of matrix behaviour. General formulae with similar structure can be developed for a matrix weakened by cavities. This can be demonstrated by employing dual potentials. The main result is presented but the full development is given elsewhere.

1. Introduction. The celebrated Hashin-Shtrikman variational structure for linear media [1, 2] was generalized to deal with any two-point statistics in [3], and was extended to nonlinear problems in a series of articles, starting with [4, 5, 6]. The present work begins with a brief review of this nonlinear extension of [1, 2] and then explores its implications for a two-component composite with arbitrary two-point statistics. Bounds (often) and 'self-consistent estimates' (always) can be developed, once a set of nonlinear algebraic equations has been solved. These equations collapse to a very simple form in the case of a nonlinear matrix reinforced by rigid inclusions, and result in the development of a closed-form approximation for the overall potential energy, for arbitrary matrix behaviour and any distribution of rigid inclusions. The formula giving this approximation is the principal result of this work. Illustrative results, for a statistically isotropic array of rigid inclusions in a particular nonlinear matrix, are presented, and certain limitations of the self-consistent prescriptions are exposed.

The case of a nonlinear matrix weakened by voids is correspondingly simple, if a formulation based on dual potentials is employed. This is mentioned briefly and a full discussion is given in [7]. The paper is concluded with an indication of work in progress.

* School of Mathematical Sciences, University of Bath, Bath BA2 7AY

2. General formulation. The composites that are discussed in this work are made of constituents, which are perfectly bonded across interfaces, whose constitutive behaviour can be expressed in terms of potential functions. Thus, for the general theory, an inhomogeneous medium, occupying a domain Ω, with potential function $W(e;\, x)$ is considered. For ease of exposition, e is regarded as the tensor of infinitesimal strain so that W represents strain energy density. An alternative interpretation of the same mathematics would be to regard e as strain-rate tensor; then, W would represent the stress potential for steady creep. The function W is taken as convex in e for each $x \in \Omega$. Following Hill [8, 9] the overall behaviour of the medium is defined from the solution of the equations of equilibrium for the medium, when it is subjected to the displacement boundary condition

$$u_i = \overline{e}_{ij}\, x_j \ , \quad x \in \partial\Omega \tag{2.1}$$

so that the six constants $\overline{e}_{ij}$ define the mean strain tensor $\overline{e}$ over Ω ($\overline{e}_{ij} = \overline{e}_{ji}$ is assumed, since asymmetry would generate in addition only an infinitesimal rigid rotation). The boundary value problem so obtained has associated with it a mean energy density $\tilde{W}(\overline{e})$ which can be characterized directly, using the minimum energy principle, as

$$\tilde{W}(\overline{e}) = \inf_{e'} \frac{1}{|\Omega|} \int_\Omega W(\overline{e} + e' \,;\, x)\, dx \ , \tag{2.2}$$

the infimum in (2.2) being taken over strain fields e' which are associated with displacement fields v which vanish on $\partial\Omega$.

When the energy function W displays a complicated variation with x, as it will for a composite, the exact calculation of $\tilde{W}(\overline{e})$ is difficult but (2.2) can be exploited to generate bounds and other stationary estimates: the simplest is obtained by setting $e' = 0$ in (2.2) to yield

$$\tilde{W}(\overline{e}) \le \frac{1}{|\Omega|} \int_\Omega W(\overline{e} \,;\, x)\, dx \ . \tag{2.3}$$

The bound (2.3) contains only 'one-point' information: it is insensitive to the spatial arrangement of the material. It is therefore of universal applicability but its generality also condemns it to be relatively crude. The present work is devoted to finding approximations for $\tilde{W}(\overline{e})$, which make allowance for 'two-point' information, in a sense made more precise later.

The tool that is employed is a generalization to nonlinear problems of the variational principle of Hashin and Shtrikman [1]. It was developed in references [4, 5, 6]. First, a 'comparison medium' is selected, whose energy function is $W_0(e)$. Then, the Legendre transform $U(\tau \,;\, x)$ of $(W - W_0)$ is introduced:

$$U(\tau \,;\, x) = \operatorname*{stat}_{e} \{\tau \cdot e - W(e \,;\, x) + W_0(e)\} \ . \tag{2.4}$$

The expression on the right side of (2.4) is stationary with respect to e, with τ fixed.

Assuming differentiability of W and W_0 with respect to e, it is attained when

$$\tau = W'(e\ ;\ x) - W_0'(e) \quad , \tag{2.5}$$

so that τ may be identified as a 'stress polarization'.

If the stationary value in (2.4) is always the supremum (as it is, for instance, if $W - W_0$ is convex), it follows that

$$W(e\ ;\ x) \geq \tau\cdot e + W_0(e) - U(\tau\ ;\ x) \tag{2.6}$$

for any e, τ and hence, from (2.2), that

$$\tilde{W}(\bar{e}) \geq \inf_{e'} \frac{1}{|\Omega|} \int_\Omega \{\tau\cdot(\bar{e} + e') + W_0(\bar{e} + e') - U(\tau\ ;\ x)\}\ dx \ , \tag{2.7}$$

for any τ . Conversely, if the stationary value in (2.4) is the infimum (for instance if $W_0 - W$ is convex), the expression on the right side of (2.7) bounds $\tilde{W}(\bar{e})$ from above. More generally, the right side of (2.7) yields an *approximation* for $\tilde{W}(\bar{e})$, for any given τ: this is written

$$\tilde{W}(\bar{e}) \approx \inf_{e'} \frac{1}{|\Omega|} \int_\Omega \{\tau\cdot(\bar{e} + e') + W_0(\bar{e} + e') - U(\tau\ ;\ x)\}\ dx \ . \tag{2.8}$$

The evaluation of the infimum in (2.8) is easy to accomplish (formally at least) if W_0 is taken as quadratic, corresponding to linear material behaviour:

$$W_0(e) = \tfrac{1}{2}\, e\cdot L_0\, e \quad , \tag{2.9}$$

where L_0 represents a constant fourth-order 'tensor of moduli'. The infimum is attained when

$$\operatorname{div}\,(L_0 e') + \operatorname{div}\,\tau = 0\ , \quad x \in \Omega \ , \tag{2.10}$$

with e' associated with zero boundary displacement. Thus, e' is defined as the solution of a linear problem and can be expressed in the form

$$e' = -\Gamma\tau \quad , \tag{2.11}$$

where Γ is a linear operator whose kernel is obtained from second derivatives of the associated Green's tensor (see, for example, [3, 10] for detailed formulae). It is easy to prove [10] that

$$\Gamma L_0 \Gamma = \Gamma$$

and hence, by substituting (2.11) into (2.8), that

$$\tilde{W}(\bar{e}) \approx \tfrac{1}{2}\,\bar{e}\cdot L_0\bar{e} + \frac{1}{|\Omega|}\int_\Omega \{\tau\cdot\bar{e} - \tfrac{1}{2}\,\tau\cdot\Gamma\tau - U(\tau\,;\,x)\}\;dx \qquad (2.12)$$

when W_0 is given by (2.9).

3. Application to a composite. Consider now a composite medium, made of n distinct constituents, or phases, so that the rth phase has energy function $W_r(e)$. Then, in the notation of the preceding section,

$$W(e\,;\,x) = \sum_{r=1}^{n} W_r(e)\,f_r(x)\ , \qquad (3.1)$$

where f_r takes the value 1 if x lies in the rth phase and zero otherwise. A corresponding choice for $\tau(x)$ is

$$\tau(x) = \sum_{r=1}^{n} \tau_r f_r(x)\ , \qquad (3.2)$$

where the τ_r are constant tensors. The relation (2.12) then yields

$$\tilde{W}(\bar{e}) \approx \tfrac{1}{2}\,\bar{e}\cdot L_0\,\bar{e} + \sum c_r\left[\tau_r\cdot\bar{e} - U_r(\tau_r)\right] - \tfrac{1}{2}\sum\sum \tau_r\cdot A_{rs}\tau_s\ , \qquad (3.3)$$

where

$$U_r(\tau_r) = \operatorname*{stat}_{e}\,\{\tau_r\cdot e - W_r(e) + W_0(e)\}\ , \qquad (3.4)$$

c_r is the volume concentration of phase r, so that

$$c_r = \frac{1}{|\Omega|}\int_\Omega f_r(x)\;dx\ , \qquad (3.5)$$

and

$$A_{rs} = \frac{1}{|\Omega|}\int_\Omega f_r(x)\,(\Gamma\,f_s)\,(x)\;dx\ . \qquad (3.6)$$

When L_0 is such that the right side of (3.3) is a bound, this can be optimised by minimizing or maximizing with respect to the parameters τ_r . More generally, a 'stationary approximation' for $\tilde{W}$ is obtained by choosing τ_r so that the right side of (3.3) is stationary. The required values of τ_r satisfy the equations

$$c_r U_r{}'(\tau_r) + \sum_s A_{rs}\,\tau_s = c_r\bar{e}\ . \qquad (3.7)$$

These equations are in general nonlinear because the functions U_r are not quadratic.

In the case of a two-phase composite ($n = 2$), the identity $f_1 + f_2 = 1$ allows A_{rs} to be expressed in terms of a single tensor P:

$$A_{rs} = c_r(\delta_{rs} - c_s)\, P \quad \text{(no summation on } r) \;, \tag{3.8}$$

where

$$c_1 c_2 P = \frac{1}{|\Omega|} \int_\Omega f_1(\Gamma f_1)\, dx \;. \tag{3.9}$$

Then, (3.3) reduces to

$$\tilde{W}(\bar{e}) \approx \tfrac{1}{2}\, \bar{e}\cdot L_0\, \bar{e} + \sum c_r \,\{\tau_r\cdot\bar{e} - U_r(\tau_r) - \tfrac{1}{2}\, \tau_r\cdot P\tau_r\} + \tfrac{1}{2} \sum\sum c_r c_s \tau_r\cdot P\tau_s \tag{3.10}$$

and (3.7) becomes

$$U_r{}'(\tau_r) + P(\tau_r - \bar{\tau}) = \bar{e} \;, \tag{3.11}$$

with $\bar{\tau} = \sum c_r \tau_r$. It should be noted that the geometrical arrangement of the composite enters these formulae through the tensors A_{rs} (or P in the case $n = 2$), so that the formulae make allowance for two-point information, in the form of weighted averages of $f_r(x)\, f_s(x')$.

Equations (3.11) have a simple explicit solution (for any P) when the composite consists of one phase, $r = 1$, with energy function $W_1(e)$, reinforced by rigid inclusions, so that

$$\begin{aligned} W_2(e) &= 0 \;, \quad e = 0 \\ &= +\infty \;, \quad e \neq 0 \;. \end{aligned} \tag{3.12}$$

The stationary value (3.4) has to be interpreted, for $U_2(\tau)$, as the supremum when W_2 is given by (3.12). Thus, $U_2(\tau) = 0$ and equations (3.11) give

$$\begin{aligned} U_1{}'(\tau_1) + c_2 P(\tau_1 - \tau_2) &= \bar{e} \;, \\ c_1 P(\tau_2 - \tau_1) &= \bar{e} \;. \end{aligned} \tag{3.13}$$

Therefore, defining

$$\gamma_1 = U_1{}'(\tau_1) \;, \tag{3.14}$$

equations (3.13) give

$$\gamma_1 = \overline{e}/c_1 \tag{3.15}$$

and so, from the inverse form of the Legendre transform (3.4),

$$\tau_1 = W_1{}'(\overline{e}/c_1) - L_0\overline{e}/c_1 \quad . \tag{3.16}$$

Then, from the second of equations (3.13),

$$\tau_2 = W_1{}'(\overline{e}/c_1) + (P^{-1} - L_0)\, \overline{e}/c_1 \quad . \tag{3.17}$$

Substituting into (3.10) then yields

$$\tilde{W}(\overline{e}) \approx \tfrac{1}{2} \frac{c_2}{c_1}\, \overline{e}\cdot(P^{-1} - L_0)\, \overline{e} + c_1 W_1(\overline{e}/c_1) \tag{3.18}$$

as a 'stationary approximation' for $\tilde{W}(\overline{e})$, in which L_0 remains to be chosen appropriately.

A lower-bound estimate for $\tilde{W}(\overline{e})$ can always be obtained by choosing $L_0 = 0$, since then $U_r(\tau)$ for $r = 1, 2$ are both defined as suprema. The tensor P increases without bound as L_0 tends to zero; correspondingly, P^{-1} tends to zero and (3.18) gives

$$\tilde{W}(\overline{e}) \geq c_1 W_1(\overline{e}/c_1) \quad . \tag{3.19}$$

This bound can also be produced directly, starting from the complementary energy principle, to which (2.7) in any case reduces when $W_0 = 0$. No non-trivial upper bound can be generated from (3.18), since $U_2(\tau)$ has to be defined as a supremum for any finite L_0 .

A lower bound better than (3.19) would be obtained if any $L_0 \neq 0$ could be found such that $U_1(\tau_1)$ is still defined as a supremum: this could be taken up in the context of particular examples.

If the requirement to generate a bound is relaxed, it is natural to attempt to identify a value of L_0 that makes the right side of (3.18) as close as possible to the exact $\tilde{W}(\overline{e})$. Since the latter is not known, the best that can be done is to offer some possible 'self-consistent' prescriptions for the estimation of $\tilde{W}(\overline{e})$. One such prescription involves the *postulate* that (3.18) would be exact if L_0 were chosen so that the mean polarization $\overline{\tau}$ is zero. This admits the interpretation that the stress in the comparison medium is equal to the mean stress in the composite; there is also a sense in which requiring $\overline{\tau} = 0$ may render the polarizations τ_1 , τ_2 as small as possible. If this prescription is implemented, it follows from equations (3.16) and (3.17) that L_0 should be chosen so that

$$(L_0 - c_2P^{-1})\, (\overline{e}/c_1) = W_1{}'(\overline{e}/c_1) \quad . \tag{3.20}$$

An alternative prescription is obtained by *postulating* that (3.18) is exact when L_0 is chosen so that the energy in the comparison medium is equal to that in the composite. This implies that L_0 should be chosen so that

$$\tfrac{1}{2}\left[\frac{\bar{e}}{c_1}\right]\cdot(L_0 - c_2P^{-1})\left[\frac{\bar{e}}{c_1}\right] = W_1(\bar{e}/c_1) \ . \tag{3.21}$$

It should be noted that (3.20) and (3.21) reduce to the same equation when W_1 is quadratic, corresponding to a linear matrix. More generally, they differ and each defines an L_0 that depends on the present level $\bar{e}$ of the mean strain. Equations (3.20), (3.21) apply for any 'two-point tensor' P and provide generalizations, to a nonlinear matrix, of the 'self-consistent QCA' as discussed, for example, in [11].

4. Example. Some calculations have been performed, for an isotropic, incompressible matrix, containing an isotropic distribution of rigid inclusions. The behaviour of the matrix is defined by an energy function W_1, which depends on the strain e only through the 'equivalent strain' $e_e = (2e_{ij}e_{ij}/3)^{1/2}$. Its constitutive relation is thus

$$\sigma_{ij} = \frac{2}{3}\left[\frac{\sigma_e}{e_e}\right] e_{ij} + \sigma_m\delta_{ij} \ , \tag{4.1}$$

where σ_m is the (indeterminate) mean stress and

$$\sigma_e = W_1{}'(e_e) \ . \tag{4.2}$$

The 'equivalent stress' σ_e is given as $\sigma_e = (3\sigma'_{ij}\sigma'_{ij}/2)^{1/2}$, where $\sigma'_{ij} = \sigma_{ij} - \sigma_m\delta_{ij}$. The definitions of σ_e and e_e are such that, under uniaxial tension σ_{33} , $\sigma_e = \sigma_{33}$ and $e_e = e_{33}$, the corresponding axial strain; thus, the relation (4.2) describes the stress-strain curve for uniaxial tension.

The form assumed for W_1 in the calculations is such that, under uniaxial tension,

$$e_e = F(\sigma_e) \ , \tag{4.3}$$

where

$$F(s) = e_0\left\{k\left[\frac{s}{\sigma_0}\right] + \left[\left[\frac{s}{\sigma_0}\right]^n - \left[\frac{\sigma_y}{\sigma_0}\right]^n\right] H(s-\sigma_y)\right\} \ , \tag{4.4}$$

with $n > 1$ and $k = 0$ or 1: the material thus provides a 'deformation theory' model of power-law hardening plasticity, rigid-plastic if $k = 0$ or elasto-plastic if $k = 1$. The special case $k = 0$, $\sigma_y = 0$ was analyzed (with the interpretation of power-law creep) by

Ponte Castañeda and Willis [12]. The function $W_1(e_e)$ associated with (4.3), (4.4) grows slower than quadratically as $e_e \to \infty$ so there is no prospect of obtaining a lower bound other than (3.19). Self-consistent estimates, based on (3.20) and (3.21), have, however, been obtained, taking the comparison material to be incompressible and isotropic, with shear modulus μ_0 .

Slightly more generally, if L_0 is isotropic and compressible, with bulk modulus κ_0 and shear modulus μ_0, it has been shown (see [3], for example) that P is isotropic and corresponds to a 'bulk modulus' κ_P and 'shear modulus' μ_P given by

$$3\kappa_P = \frac{1}{3\kappa_0 + 4\mu_0}, \quad 2\mu_P = \frac{3(\kappa_0 + 2\mu_0)}{5\mu_0(3\kappa_0 + 4\mu_0)} \quad . \tag{4.5}$$

The tensor P^{-1} then corresponds to a 'bulk modulus' $1/(9\kappa_P)$ and 'shear modulus' $1/(4\mu_P)$. In the incompressible limit, $\kappa_0 \to \infty$ and the 'bulk modulus' behaviour of P^{-1} becomes singular. However, the mean strain $\bar{e}$ contains no dilatation so only the 'shear modulus' associated with P^{-1} appears in (3.20) or (3.21).

The general formula (3.18) reduces to

$$\tilde{W}(\bar{e}) \approx \frac{c_2}{c_1}\frac{9}{4}\mu_0(\bar{e}_e)^2 + c_1 W_1(\bar{e}_e/c_1) \quad , \tag{4.6}$$

while

$$W_0(e) = \frac{3}{2}\mu_0(e_e)^2 \quad . \tag{4.7}$$

Considering first the prescription (3.20) (which is equivalent to $W_0{}' = \tilde{W}'$), this gives

$$\mu_0 = \frac{2c_1}{3(2 - 5c_2)}\frac{W_1{}'(\bar{e}_e/c_1)}{\bar{e}_e} \tag{4.8}$$

and this μ_0, when substituted into (4.6), yields the self-consistent estimate

$$\tilde{W}_{(1)}(\bar{e}) = \frac{3c_2}{2(2 - 5c_2)}\bar{e}_e W_1{}'(\bar{e}_e/c_1) + c_1 W_1(\bar{e}_e/c_1) \quad . \tag{4.9}$$

The prescription (3.21) ($W_0 = \tilde{W}$) gives the estimate

$$\tilde{W}_{(2)}(\bar{e}) = \frac{2c_1^2}{(2 - 5c_2)} W_1(\bar{e}/c_1) \quad . \tag{4.10}$$

The estimates (4.9) and (4.10) coincide when W_1 is quadratic, corresponding to linear behaviour. They both become singular at $c_2 = 0.4$ and hence must be regarded with caution at high values of c_2; the corresponding estimate for a linear matrix shares this feature.

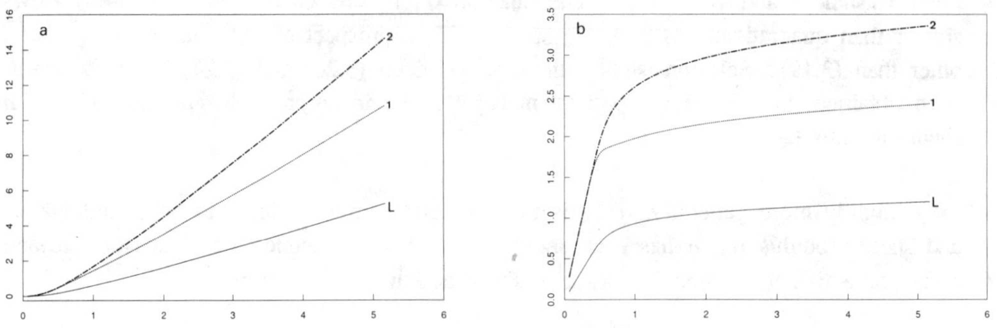

Figure 1. Estimates for (a) $\tilde{W}/(\sigma_0 e_0)$ versus $\bar{e}/e_0$ and (b) $\bar{\sigma}_e/\sigma_0$ versus $\bar{e}/e_0$, for a composite with $c_2 = 0.3$, $n = 10$, $k = 1$. L: lower bound, 1: self-consistent estimate $\tilde{W}_{(1)}$, 2: self-consistent estimate $\tilde{W}_{(2)}$.

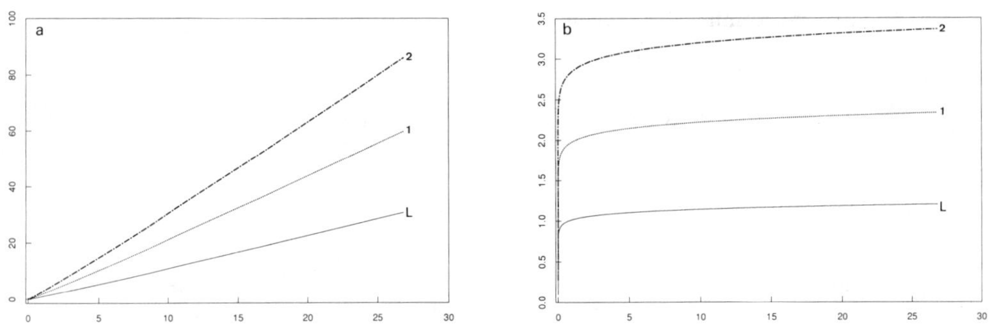

Figure 2. Plots as for Fig. 1 except that $c_2 = 0.3$, $n = 20$, $k = 0.01$.

All of the results that are displayed took $\sigma_y/\sigma_0 = 0.2$. Figure 1a shows plots of the lower bound (3.19) (labelled L) and the self-consistent estimates (4.9) and (4.10) (labelled 1 and 2), in the form of $\tilde{W}/(\sigma_0 e_0)$ versus $\bar{e}_e/e_0$, for the parameter values $c_2 = 0.3$, $n = 10$ and $k = 1$. Figure 1b shows plots of the three corresponding uniaxial stress-strain relations. The one associated with the bound (3.19) is the lowest though it is not guaranteed to be a lower bound beyond the linear regime. Figure 2a shows plots similar to those in Fig. 1a, except that $c_2 = 0.3$, $n = 20$ and $k = 0.01$ (the program as written required $k > 0$) and Fig. 2b shows the corresponding uniaxial stress-strain curves.

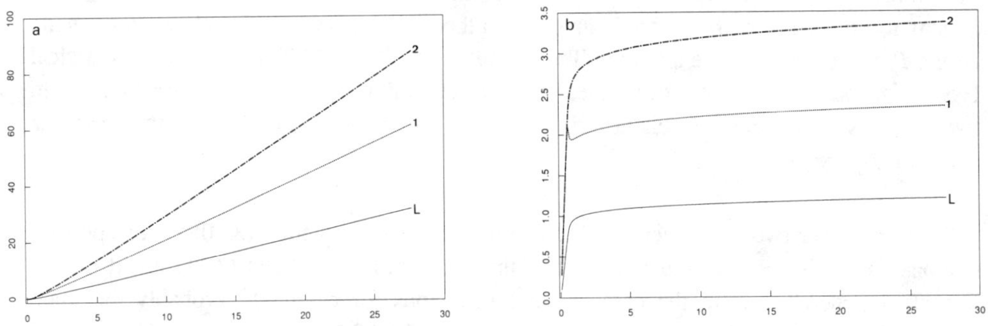

Figure 3. Plots as for Fig 1. except that $c_2 = 0.3$, $n = 20$, $k = 1$.

Figures 3a and 3b are similar, except that $c_2 = 0.3$, $n = 20$ and $k = 1$. The noteworthy feature is that the self-consistent estimate $\tilde{W}_{(1)}$ is not convex: this is brought out in Fig. 3b, which displays a non-monotonic stress-strain curve. It can be proved [12] that the exact overall energy function $\tilde{W}(\bar{e})$ is convex and this property is shared by the approximation (4.6), with μ_0 fixed. The problem occurs once μ_0 is allowed to vary with $\bar{e}$.

5. Discussion. Ponte Castañeda and Willis [12] studied the overall behaviour of an incompressible matrix displaying pure power-law creep, reinforced by rigid inclusions or weakened by voids. Dimensional analysis dictates that the overall energy function should be homogeneous, of degree $(n + 1)/n$. The direct approximation, (4.6) for the case of rigid inclusions, does not have this property, though the self-consistent estimates (4.9) and (4.10) do. Thus, the self-consistent estimates do have some desirable features but the failure of convexity noted above militates against (4.9). This particular failure occurs when elasticity is present ($k = 1$) and n is large; it was not found by Ponte Castañeda and Willis [12] in the case of pure power-law behaviour. Failure of convexity was, however, noted for a pure power-law matrix weakened by cavities; furthermore, the estimate corresponding to (4.10) violated a bound under some circumstances. Further discussion of the problem of a matrix containing cavities has been given by Willis [7], with examples given for matrix behaviour defined by (4.3) and (4.4). This work cannot be reviewed in detail except to mention that the general formulation can be made to look exactly like that given in Section 3, by working in terms of dual potentials $W_r{}^*(\sigma)$, defined so that

$$W_r{}^*(\sigma) = \sup_e \{\sigma \cdot e - W_r(e)\} \quad . \tag{5.1}$$

The energy function W_2 for a cavity is identically zero, so $W_2{}^*$ has a form like (3.12), the variable σ replacing e. Some detailed analysis is required to prove this, but the result corresponding to (3.18), for a matrix weakened by cavities, is

$$\tilde{W}^*(\bar{\sigma}) \approx \tfrac{1}{2} \frac{c_2}{c_1} \, \bar{\sigma} \cdot (P^{-1} - L_0)^{-1} \, \bar{\sigma} + c_1 W_1{}^*(\bar{\sigma}/c_1) \quad . \tag{5.2}$$

The formulae (3.18) and (5.2) allow for two-point information through the tensor P . No assumption on the microgeometry is involved, although some assumptions have to be made before P can be evaluated explicitly. The formulae (3.18) and (5.2) generalize the classical Hashin-Shtrikman estimates for a linear matrix containing rigid inclusions or cavities, in two respects: the matrix is characterized by any convex function W_1 and the formulae apply to any two-point statistics.

If any other two-component medium were considered, equation (3.10) yields approximations for $\tilde{W}$ but the solution of the non-linear algebraic equations (3.11) could not be expressed generally, as in the case of rigid inclusions (or cavities). Probably the main source of inaccuracy of the approximations developed in this work is the combination of using a linear comparison medium and very simple polarizations resulting inevitably in strain fields that cannot be exactly realized in a nonlinear medium. One possible remedy which is under consideration at present is to adopt an approach based on conditional averaging, as developed for linear media by Willis [11]. In this approach, a linear comparison medium is still used but integral equations for the polarizations are developed: these are nonlinear in the case of a nonlinear composite and allow accurate modelling of the strain around an inclusion. The penalties include the sacrifice of a variational structure and the need to solve nonlinear equations. This work will be reported elsewhere.

REFERENCES

[1] Z. Hashin and S. Shtrikman, *On some variational principles in anisotropic and non-homogeneous elasticity*, J. Mech. Phys. Solids, **10** (1962), pp. 335-342.

[2] Z. Hashin and S. Shtrikman, *A variational approach to the theory of the elastic behaviour of multiphase materials*, J. Mech. Phys. Solids, **11** (1963), pp. 127-140.

[3] J. R. Willis, *Bounds and self-consistent estimates for the overall moduli of anisotropic composites*, J. Mech. Phys. Solids, **25** (1977), pp. 185-202.

[4] J. R. Willis, *The overall elastic response of composite materials*, J. Appl. Mech., **50** (1983), pp. 1202-1209.

[5] J. R. Willis, *Variational estimates for the overall response of an inhomogeneous non-linear dielectric*, in Homogenization and Effective Moduli of Materials and Media, J. L. Eriksen, D. Kinderlehrer, R. Kohn and J.-L. Lions, eds., Springer-Verlag, New York, 1986, pp. 245-263.

[6] D. R. S. Talbot and J. R. Willis, *Variational principles for inhomogeneous nonlinear media*, IMA J. Appl. Math., **35** (1985), pp. 39-54.

[7] J. R. Willis, *Variational estimates for the overall behaviour of a nonlinear matrix-inclusion composite*, in Micromechanics and Inhomogeneity - The Toshio Mura Anniversary Volume, G. J. Weng, M. Taya and H. Abe, eds., Springer-Verlag, New York, 1989 (to appear).

[8] R. Hill, *Elastic properties of reinforced solids: some theoretical principles*, J. Mech. Phys. Solids, **11** (1963), pp. 357-372.

[9] R. Hill, *On constitutive macro-variables for heterogeneous solids at finite strain*, Proc. Roy. Soc. London, **A326** (1972), pp. 131-147.

[10] J. R. Willis, *Variational and related methods for the overall properties of composites*, in Advances in Applied Mechanics, Volume 21, C.-S. Yih, ed., Academic Press, New York, 1981, pp. 1-78.

[11] J. R. Willis, *Some remarks on the application of the QCA to the determination of the overall response of a matrix/inclusion composite*, J. Math. Phys., **25** (1984), pp. 2116-2120.

[12] P. Ponte Castañeda and J. R. Willis, *On the overall response of nonlinearly viscous composites*, Proc. Roy. Soc. London, **A416** (1988), pp. 217-244.

PART II

Network Models

CHAPTER 8

Scaling Theory and Fluctuations of the Strength of Composites

P. M. Duxbury*
Y. S. Li*

Abstract. An introduction to the statistical mechanics of breakdown in random media is given. Electrical, dielectric and mechanical breakdown are studied in a quenched random environment, and most discussion centers on random microstructures obeying lattice percolation statistics. The strength of such percolation networks exhibits a "dilute limit catastrophe" in which any finite fraction of disorder drastically reduces the network strength. This is in contrast to transport or elastic moduli which are linear in dilution with slope O(1) near the pure limit. The origin of this difference is explained by the fact that strength depends on extreme fluctuations, while moduli depend on an average over all fluctuations. One important consequence is that for $p \neq p_c$, sample to sample fluctuations in moduli are $O(1/(V^{1/2}))$, while sample to sample fluctuations in strength are shown to be $O(1/(\ln V)^2)$. Near the percolation point critical behavior occurs, and breakdown critical exponents are presented. The simplest breakdown networks are brittle, and generalizations that exhibit toughening are discussed.

I. Introduction. Breakdown strength, either electrical or mechanical, is a property of prime importance in most applications of composites. Prediction of the effect of composite microstructure on

* Department of Physics and Astronomy and Center for Fundamental Materials Research, Michigan State University, East Lansing, MI 48824-1116 U.S.A.

breakdown strength is however, at a far more primitive stage than corresponding predictions of effective conductivity or elastic moduli. Recent generalizations of the percolation model to study breakdown phenomena[1-7] provide a new starting point in this class of problem, and an important pedagogical system in which to test the relative importance of various microstructural features and randomness on breakdown phenomena. In this paper we give an introduction and overview of the ideas and results that have been recently derived using this class of model. In addition these models are placed in the broader context of other statistical and statistical mechanics approaches to breakdown in random media.

To be specific, we consider the failure of a spring, the triggering of a circuit-breaker, the failure of a piece of dielectric and the superconducting to normal transition to be important illustrative examples of breakdown phenomena (see Fig. 1).

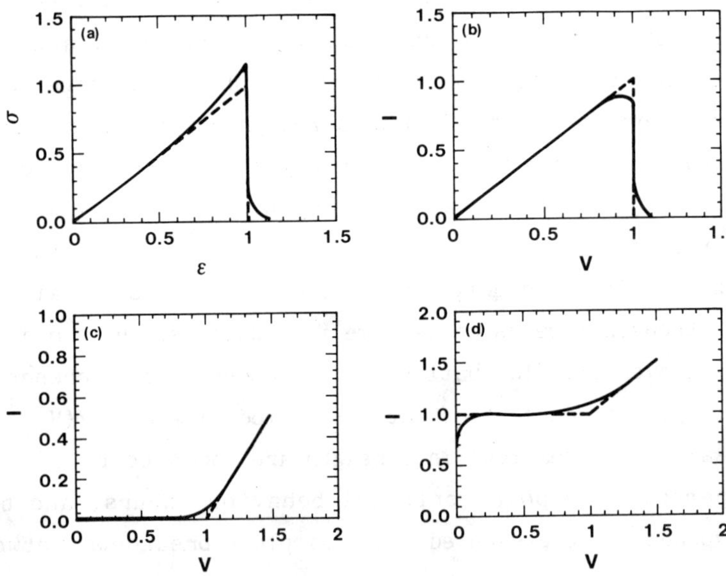

Figure 1 Examples of breakdown phenomena a) The failure of a spring; b) The failure of a fuse; c) The failure of a dielectric and; d) A superconductor characteristic. In all of these figures, the dotted line is a discontinuous idealization of the more realistic (solid line) characteristics.

To a first approximation these example show linear response up to a threshold load at which they show a discontinuity in their response (dotted line in Fig. 1). If a disordered system shows this sort of response locally the macroscopic response is a complicated function of the external load level, the precise nature of the local characteristics, system size and disorder. Disorder effects are especially interesting and difficult to study, as local failure in weak regions or regions carrying large loads may lead to load redistribution throughout the material. Understanding of breakdown phenomena in random media then relies on an understanding of the initial (intrinsic) load distribution, the criterion for local failure, and the detailed process of load redistribution after local failure.

Most traditional theories of mechanical fracture study single cracks using expanded zones near the crack tip to incorporate intrinsic defects or non-linearities. Percolation models, and statistical mechanics models in general, are used to study systems with random microstructure and to understand the relative importance of local as compared to non-local load redistribution effects caused by local failure.

The paper is arranged as follows. In Section II we compare breakdown scaling with moduli scaling by first considering a single crack, and then using percolation models as typical examples. One striking example of the marked differences between these two classes of properties is that in the dilute flaw limit, the breakdown strength of percolation models shows a singular scaling behavior[4,5], while moduli do not[9-11]. The origin of this difference is clearly demonstrated if one considers the distribution of local stresses or electric fields. Transport and elastic moduli are related to the second moment of these distributions, while breakdown is related to the very high (infinite) moments. In the case of especially sharp flaws (cracks), even relatively low moments ($m>4$) have a significantly different scaling behavior than the second moment[11,12].

Another implication of the dependence of breakdown properties on the tail of the local load distribution is that sample to sample variations in strength conform to extreme statistics distributions[13,14], rather than the central limit form appropriate to the conductivity or elastic moduli. The particular reliability distribution relevant to percolation models of breakdown[5,15] is discussed in Section III.

Near percolation points, breakdown critical exponents occur, and various expressions and approximate bounds on these exponents are given in Section IV.

In Section V, we note that the simplest percolation models of breakdown describe brittle failure. Extensions that model tough composites are discussed[16], with emphasis on the ligamentary (residual) bonding mechanism common in ceramic and fiber composite failure.

Section VI concludes the paper with an overview and perspective of the directions in which this new field is currently evolving.

II. COMPARISON OF MODULI AND BREAKDOWN SCALING.

Isolated Crack-like Flaw in a Homogeneous Background. Consider an isolated crack of length, a, with a curvature at the crack tip, κ, in an otherwise homogeneous conductor with large volume $V = L^d$ ($a \ll L$). It is known that such an isolated crack makes negligible contribution to the overall conductivity, Σ_{eff} i.e. that,

$$\Sigma_{eff} \sim \Sigma_0 - O(1/V) \tag{1}$$

where Σ_0 is the background conductivity. By contrast, this crack drastically reduces the electrical and mechanical strength of the system, as near the crack tip, strong load enhancements occur, and for the electrical case one finds[7],

$$E_{tip}/E_0 \sim (a/2\kappa)^{1/2} \tag{2}$$

which leads to an electrical strength of order (assuming unit strength in the pure limit)

$$E_b(a)/E_b(0) \sim E_0/E_{tip} \sim (2\kappa/a)^{1/2} \tag{3}$$

Where $E_b(a)$ is the electrical strength of the sample with a flaw in it, and $E_b(0)$ is the electrical strength of the material with no flaws, and the electric field is applied perpendicular to the long axis of the crack. Similar expressions to (2) and (3) hold for the tensile strength of brittle materials with E_b and E_0 replaced by stresses, and with a different constant prefactor. The important scaling behavior that we want to emphasize is that in many materials electrical and mechanical strength decreases as the square root of crack length, while conductivity and elastic moduli have corrections of order $(1/V)$ for $a \ll L$. The results (2) and (3) are based on field intensity calculations. Fracture strength reduction due to cracks is most often discussed using energy balance arguments, in which the elastic energy reduction due to increasing crack length is balanced against the energy to form new fracture surfaces. For the linear elastic case this is a classical calculation, and leads to the (plain stress) Griffith formula[17],

$$2Y = \pi \sigma_b^2 a/e \quad \text{which implies} \quad \sigma_b = (2Ye/\pi a)^{1/2} \tag{4}$$

where e is the elastic modulus of the background and Y is the energy per unit length required to create new crack surface. Although it is not as well known, an analogous result[18] is available for the dielectric strength of a two dimensional insulator containing a long thin piece of metal (this is a dielectric "crack") oriented parallel to the direction of applied electric field. The result is

$$4G_e = \pi E_b^2 a\varepsilon \quad \text{which implies } E_b = (4G_e/\varepsilon\pi a)^{1/2} \quad (\text{only for 2-d})^{18} \tag{5}$$

where ε is the dielectric constant of the background, and G_e is the energy required to induce ionization of enough atoms to cause unit crack extension. Note however that this analogy does not carry over to 3 dimensions where a penny shaped crack is most important for mechanical and electrical failure, while a finger-like defect is most important for dielectric failure. However, both results (4) and (5) confirm the square root reduction in breakdown strength with crack length predicted from field intensity calculations.

The crossover between the low moment scaling behavior typified by equation (1), and high moment scaling behavior typified by equations (2) to (5) is reflected in the scaling behavior of the complete moment spectrum of the stress or electric field occurring in the system[11]. To demonstrate this note that the maximum electric field occurring in a system may be written,

$$E_{max} \sim \langle E^m \rangle^{1/m}{}_{m\to\infty} \tag{6a}$$

so that

$$E_b/E_0 \sim \langle E_0\rangle/\langle E^m\rangle^{1/m} \quad \text{as } m\to\infty \tag{6b}$$

For the case of insulating inclusions, it can be seen that the effective conductivity is related to the second moment of the electric field distribution by noting

$$\text{Power} = V\,E_0{}^2\Sigma_{eff} = \int E^2\Sigma(\bar{r})d\bar{r} = \Sigma_0\,\langle E^2\rangle\,V \tag{7}$$

so that

$$\Sigma_{eff}/\Sigma_0 = \langle E_0{}^2\rangle/\langle E^2\rangle \tag{8}$$

We are then lead to believe that the function

$$R_m = \langle E_0\rangle/\langle E^m\rangle^{1/m} \tag{9}$$

in a system containing an isolated crack, should show a crossover from O(1/V) corrections when m is small to the typical breakdown scaling of equation (3) when m becomes large. To study this crossover analytically consider the expression

$$\langle\, E^{\,m}\rangle = (1/V)\int_V (E_\xi{}^2 + E_\eta{}^2)^{m/2}\,dV \tag{10}$$

Where ξ and η are two dimensional elliptic co-ordinates. We have not been able to analytically evaluate equation (10) for any $m>2$, but the important scaling behavior may be found by noting that the electric field as a function of radial distance parallel to the major axis of the ellipse has the form;

$$E_z(r)/E_0 \sim \begin{cases} 1 + \alpha_{2e}/r^d & \text{for } r \gg a \quad (11a)\\ c_{1e} + c_{2e}(a/2r)^{1/2} & \text{for } \kappa < r < a \quad (11b)\\ k_{1e} + k_{2e}(a/2\kappa)^{1/2} & \text{for } r < \kappa \quad (11c)\end{cases}$$

Where α_{2e} is a dipole moment for the two dimensional problem of a crack in a resistive background, κ is the crack tip curvature ($=b^2/a$) for an ellipse and c_{1e}, c_{2e}, k_{1e} and k_{2e} are constants for the two dimensional electrical problem. In fact, the scaling behavior of equation (11) is highly universal, and applies (with different α, c and k) to ellipse, ellipsoidal and slit-like cracks in 2-d and 3-d linear elastic (plain stress) and electric problems (see ref. 11). The most important physical observation is that as the moment value is increased, the regions (11b) and (11c) increase their contribution to the integral (10). This is explicitly seen by subdividing and approximating the integral (10) as shown below,

$$V \langle [E(r)/E_0]^m \rangle \sim \int_{r>a} r^{d-1} dr \, (1 + \alpha_{2e}/r^d)^m \quad +$$

$$\int_{\kappa<r<a} rdr \, (c_{1e} + c_{2e}(a/2r)^{1/2})^m \; + \int_{r<\kappa} rdr \, (k_{1e} + k_{2e}(a/2\kappa)^{1/2})^m \qquad (12)$$

keeping the most singular terms in each integral, we find;

$$\langle [E(r)/E_0]^m \rangle \sim 1 + m \, O(V_{inc}/V) + (a/2\kappa)^{m/2} \, \kappa^2/V \qquad (13)$$

Where V_{inc} is the volume of the inclusion. The $O(V_{inc}/V)$ term is estimated by considering the leading term in the first integral, and including the correct angular terms. This is necessary, as this term is a dipole integral, in which care must be taken to ensure that all terms are included, and that boundary terms do not produce unphysical results. The result is a term of order the inclusion energy, which leads directly to the result quoted.

From equation (13), we can see that R_m (see equation (9)) has the following scaling forms

$$R_m \sim \begin{cases} 1 - O(V_{inc}/V) & m \ll m_c \\ (2\kappa/a)^{1/2} & m \gg m_c \end{cases} \qquad (14)$$

where m_c is found from

$$(a/2\kappa)^{m/2} \kappa^2/V \sim 1 \qquad (15)$$

or

$$m_c \sim 2\ln(V/\kappa^2)/\ln(a/2\kappa) \qquad (16)$$

Due to the universality of equation (11), this result straightforwardly generalizes to three dimensional electric problems and to elastic problems in two and three dimensions[11].

There is also a great deal of similarity between the problem of an insulating crack in an conducting background and the problem of a conducting crack in an insulating background. The former problem has been discussed as the simplest example of an electrical breakdown problem[5], while the latter is the simplest starting point for understanding defect induced dielectric breakdown[5,9]. It is well known that in two dimensions, the electric field enhancement at the tip of a metal ellipse oriented parallel to the applied electric field in an insulating background, is the same as that at the tip of a void inclusion oriented perpendicular to the applied field and in a conducting background. The asymptotic form (11) then applies equally well to the dielectric problem in two dimensions, and hence that equations (13) to (16) express the moment scaling appropriate to that problem. The dielectric problem is different in three dimensions however, as there the most important crack-like defect is a finger-like inclusion[5,9], which induces the following electric field behavior near its tip;

$$E_z(r)/E_0 \sim \begin{cases} 1 + \alpha_{3d}/r^3 & \text{for } r>a \\ \ln(r/2a) + a/2r & \text{for } \kappa<r<a \\ 1 + (a/b)^2/\ln(2a/b) & \text{for } r<\kappa \end{cases} \tag{17}$$

in this case the integration over the region near the crack tip leads to,

$$V \langle [E(r)/E_0]^m \rangle \sim \int_a^L r^2(1 + \alpha_{3d}/r^3)^m \, dr + \int_\kappa^a r^2 dr \, (\ln(r/2a) + a/2r)^m + \int_\kappa^a r^2 dr \, (1 + (a/b)^2/\ln(2a/b))^m \tag{18}$$

so that

$$\langle [E(r)/E_0]^m \rangle \sim 1 + mO(V_{inc}/V) + (a/2\kappa)^m \kappa^3/V \tag{19}$$

The moment value at which crossover between low moment scaling and extreme scaling occurs is then given by,

$$m_c \sim \ln(V/\kappa^3)/\ln(a/2\kappa) \tag{20}$$

Extremely sharp cracks ($\kappa \to 0$). It is seen from equations (13) and (19) that if κ becomes sufficiently small, a singular behavior can occur for low moment values. For the problems described by equation (13) this moment value is

$$m_{c1} = 4 \tag{21}$$

while for the 3-d dielectric problem (see equation (19)),

$$m_{c1} = 3 \tag{22}$$

For $m > m_{c1}$, moments are affected by extremely sharp cracks, and in fact these moments are singular as $\kappa \to 0$.

Percolation Models of Breakdown. Percolation networks can be used to study breakdown in random media, if the present bonds (with volume fraction p) are assigned characteristics like that shown in Figure 1. If an external load is applied to such a percolation network, local failure may occur. Due to this local failure, the applied load is redistributed among the remaining unfailed bonds. Some of these bonds may then fail, and the sequence is continued until either the whole network fails, or it reaches a stable configuration. Clearly, the strength of the percolation networks depends on the local failure criterion, and the particular law used to redistribute the load after a bond has failed. The local failure criterion used in most modelling thus far is that a bond fails as shown in Fig. 1 when a stress or electric field greater than a threshold value is applied across it. In electrical networks the voltage distribution in the network with the removed bond is then found using Laplace's equation[1,2,6,19], while in mechanical networks several possible redistribution laws have been used. Central force networks[1,4], networks with bond bending forces[14] and the Born model[15] have been studied in this context.

The network response is found by slowly increasing the external voltage or strain and following the bond breaking procedure outlined in the previous paragraph. This procedure corresponds to a very slow increase in applied load, so that at any time the system is at steady state. The network I-V characteristic of a 70x70 "fuse network" at p=0.75 found using this procedure is shown in Fig. 2a. Similarly, in

Fig. 2b, a network characteristic for the central force spring model is presented (from ref. 7).

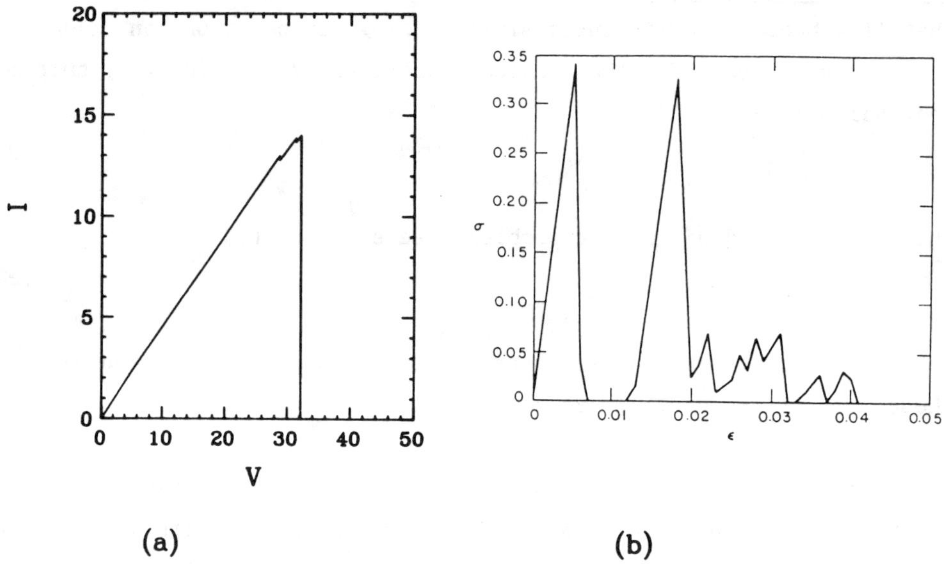

Figure 2 Random network characteristics a) The fuse network at p=0.75 and; b) The central force spring network at p=0.90

In the latter case, the network loses rigidity very soon after the first bond fails, and we consider the network to have "failed" at this point. During the evolution of the characteristics of Fig. 2, a crack grows through the random network, and it's path becomes very tortuous near the percolation point. In this paper, we choose not to address the question of crack structure and its dependence on the microscopic growth laws, although it is a topic of broad importance (see overview in Section VI). Instead, we emphasize the strength and strength statistics of breakdown networks an area where considerable analytic progress is possible.

The Strength of Percolation Networks[4,5]. The breakdown stress or electric current is given by the peak of the I-V or σ-ε characteristics of Fig. 2. It is evident from this Figure, that this peak lies very close to the point at which the first local failure occurs in the network. For this reason, these networks are considered to be brittle. A scaling analysis of the strength of these networks then reduces to a study of the scaling behavior of the peak of the appropriate characteristics, which in turn reduces (in first approximation) to a study of the point at which the first local failure occurs in the network. Most of the numerical analysis of <u>brittle</u> percolation networks has relied on these statements. Analytic arguments that quantitatively support these ideas are as follows.

Consider an LxL square lattice fuse network for illustration purposes, and note that failure of this network when p=1 is given by

$$I_b(p=1) = I_1(p=1) = L \tag{23}$$

so we define a breakdown current density;

$$j_b(p=1) = j_1(p=1) = 1 \tag{24}$$

The subscripts 1 and b refer to the external current required to induce the first local failure, and the current required to induce failure of the network respectively. Since failure is initiated in regions with the largest load, we look for local defect structures that lead to large stress or current enhancements. Rows of adjacent removed flaws (cracks) clearly do this, as is seen from equation (2). A first guess is then that the largest local load occurs at the end of the <u>largest sequence</u> of adjacent removed bonds. This crack-like void has size c given approximately by,

$$V\ (1-p)^c \sim 1 \tag{25}$$

which implies

$$c \sim -d\ln L/\ln(1-p) \tag{26}$$

The load enhancement at the tip of this defect goes as the square root of the crack length, which predicts that the largest current or stress (in the small 1-p limit) in the network scales as

$$j_{max} \sim j_0(\ 1 + k_e[-d\ln L/\ln(1-p)]^{1/2}) \tag{27}$$

where k_e is a constant that depends on the nature of the crack tip. A similar expression applies to the largest stress in a spring network, with $j_{max} \to \sigma_{max}$, $j_0 \to \sigma_0$ and $k_e \to k_m$. Since the bond carrying this large load lies at the end of a row of defects, when it fails, it makes the defect more severe, so that failure propagates from this crack and the first local failure and the network failure occur at the same external load. There are however bonds that carry larger load than that occurring at the end of the largest crack-like flaw, and one example is two adjacent crack-like flaws with an intact region between them. If the intact region is small compared to the size c of the two neighboring cracks, it may be shown in both the mechanical and electrical cases that[4,5]

$$j_{max}/j_0 \sim \sigma_{max}/\sigma_0 \sim 1 + Bc \tag{28}$$

The important fact being that the behavior is linear in c instead of a square root. Based on simple local conservation of field lines it is seen that this is the most severe field enhancement law that can apply. Since strength is proportional to the reciprocal of the largest local fields, we conclude that the strength of dilute percolation networks scales as:

$$j_b \sim j_0/[\ 1 + k_e(\ln V_n/(-\ln f))^{\alpha}] \quad \text{for electrical systems,} \tag{29}$$

$$\sigma_b \sim \sigma_0/[\ 1 + k_m(\ln V_n/(-\ln f))^{\alpha}) \quad \text{for mechanical networks} \tag{30}$$

and

$$E_b \sim E_0/[\ 1 + k_d(\ln V_n/(-\ln f))^{\beta}) \quad \text{for dielectric breakdown} \tag{31}$$

with

$$1/(2(d-1)) < \alpha < 1 \quad \text{and} \quad 1/2 < \beta < 1 \tag{32}$$

In these expressions, $V_n = L^d$ and f=1-p for percolation networks. The expressions may be extended to experimental composites containing voids (with a narrow size distribution) if V_n = Volume of sample/(Average void volume). For the dielectric breakdown case, the voids are replaced by metal circles or spheres. The different dimensional dependence in the lower bound in the dielectric case is due to the fact that one dimensional conducting cracks cause strong electric field enhancements in insulators, while d-1 dimensional crack-like voids are required to cause strong field enhancements in elastic or conducting backgrounds. The constants k_e, k_m and k_d are independent of f or volume, but may dependent on the elastic constants or conductivity of the background material.

One striking feature of equations (29) to (31) is that breakdown strength is singular on approach to the pure limit (clearly, no such effects occur in transport or elastic moduli). We can illustrate the origin of the difference between breakdown scaling and moduli scaling, especially in the pure limit, by considering the distribution of bond currents occurring in a percolation network. For a square lattice electrical network with a constant applied current, the distribution of bond currents is shown in Fig. 3a.

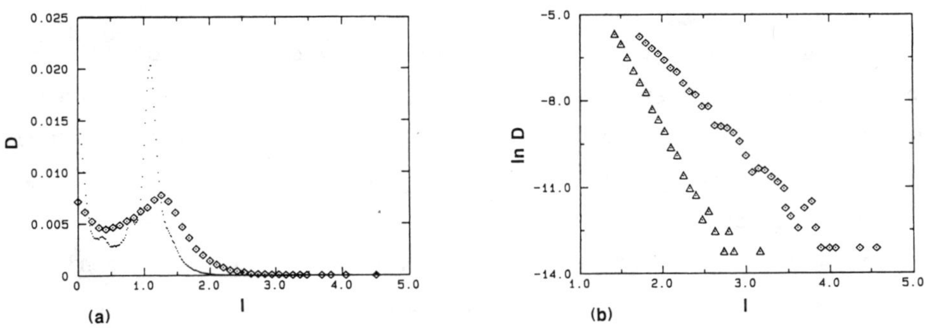

Figure 3 a) The bond current distribution of 80x80 random resistor networks (RRN); p=0.90 (o), p=0.80 (<>). b) The same data on a log-linear plot to demonstrate the exponential tail of the distribution.

We can relate the conductivity of the network to this distribution by noting that the power dissipated in the network is given by:

$$\text{Power} = \sum I_{bond}^2 / \Sigma_{bond} = I_{applied}^2 / \Sigma_{eff} \qquad (33)$$

so that

$$\Sigma_{eff} \sim \langle I_{bond}^2 \rangle^{-1} \qquad (34)$$

which is the second moment of the local load distribution plotted in Fig. 3. Since breakdown is initiated in the bond carrying the largest current, we can say that for brittle networks,

$$I_b \sim 1/I_{max} \sim \langle I^m_{bond}\rangle^{-1/m} \quad \text{in the limit } m\to\infty \tag{35}$$

as stated in electrical field language in equation (10). It is again clear that breakdown properties are related to the very high moments of the local load distribution. Since the tail of the local load distribution is exponential away from the percolation point[5] (see Fig. 3b), low moments ($m<m_c$) are dominated by the peaks in the load distribution. However for sufficiently high moments ($m>m_c$), the tail of the distribution becomes of most importance. Due to the exponential nature of the load distribution tail for $p>p_c$, this critical value of m is given by[11]

$$m_c \sim \ln L \tag{36}$$

The scaling behavior of properties dominated by the tails of distributions are very different from those dominated by the most likely or lower percentiles of these distributions. Most theoretically studied properties exhibit <u>low moment scaling</u> (e.g. conductivity or elastic moduli), percolation models of breakdown are of special interest as they provide a paradigm of properties that exhibit <u>extreme moment scaling</u>[11]. The statistical language used to study extreme properties is the <u>statistics of extremes</u>, which is used extensively in reliability analysis[13,14]. We will return to a more detailed discussion of the use of extreme statistics analysis in percolation models of breakdown in the next Section.

Moment scaling arguments such as those leading to equations (33) - (35) may be straightforwardly generalized to mechanical and dielectric problems[11]. As an illustrative example of the difference between moduli and breakdown scaling, especially in the pure limit, consider Fig. 4, in which the conductivity and electrical strength of a 100x100 square lattice percolation network is plotted as a function of disorder. The strong <u>dilute limit singularity</u> evident in Fig. 4 for the breakdown strength becomes stronger (albeit logarithmically) as the system size is increased (see equations (29) - (31)). This effect should then be clearly present in well controlled experiments, namely: mechanical strength as a function of void fraction in brittle materials, dielectric strength in insulators containing metal inclusions and electrical strength of conductors containing voids.

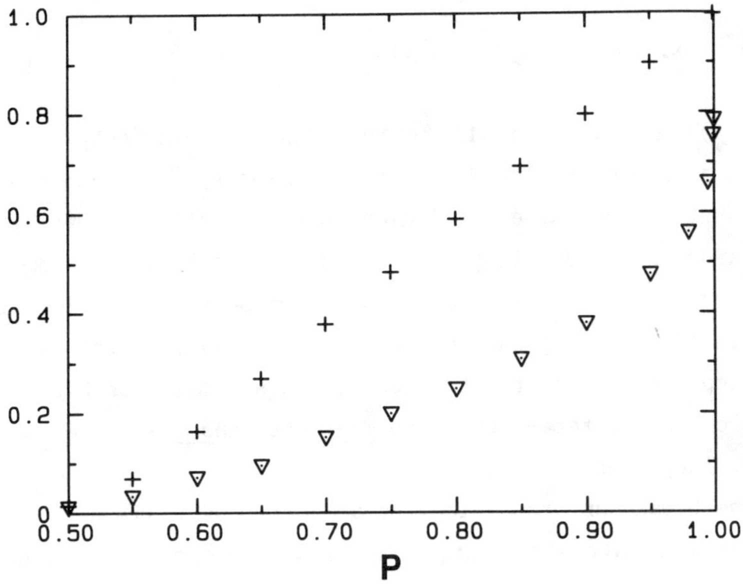

Figure 4 Plots of the conductance (+) and the breakdown current (∇) as a function of dilution for 100x100 RRN. Each point is an average over 50 configurations.

Two papers reporting the second of these experiments have recently been published[24,25], with clear indication of a rapid drop in strength for small metal volume fractions and in good agreement with equation (31). Some evidence of this effect in porous ceramics is also present in the literature[26]. There is room however for better controlled experiments on these systems, and more quantitative tests of the dilute limit scaling laws given in equations (29) - (31).

III. **STRENGTH FLUCTUATIONS AND DISTRIBUTIONS**[4,15]. The average conductivity or elastic moduli varies on lengths $L<\xi$, but saturates on lengths $L>\xi$, where ξ is the correlation length in percolation models, and a length scale over which the system is disordered in general. Since regions on lengths $L>\xi$ are uncorrelated, the average value for the moduli measured on samples of size L^d have fluctuations of order $(L/\xi)^{-d/2}$. This qualitative argument is valid for $L>>\xi$, but must be modified for $L\sim\xi$. By similar reasoning, the central limit theorem shows that the full distribution function for such a moduli and $L>>\xi$ is Gaussian, with

$$P(\Sigma_N) \sim \exp\left[-(\Sigma_N - \Sigma_\infty)^2/(Nc_e(p))\right] \tag{37}$$

where $P(\Sigma_N)$ is the probability that the conductivity of a sample measured on a sample of size $N = L^d$ has a value Σ_N. $c_e(p) \sim C_e/\xi^d$, and c_e is a constant for the electrical problem. Similar expressions apply to capacitance in dielectric problems and elastic moduli in mechanical systems, and c_e is changed to c_d and c_m respectively. Since the distribution in (37) is very narrow (for N large away from the percolation point), sample to sample fluctuations are also small for $L>>\xi$. This is the origin of the statement that, <u>conductivity and elastic moduli self-average</u> for large enough system sizes.

Because breakdown properties are dominated by extreme events, the discussion given above no longer applies. In particular, for a fixed amount of disorder, sample to sample fluctuations in strength are far larger than those for conductivity or elastic moduli. This is explicitly demonstrated for the electrical case in Fig. 5,

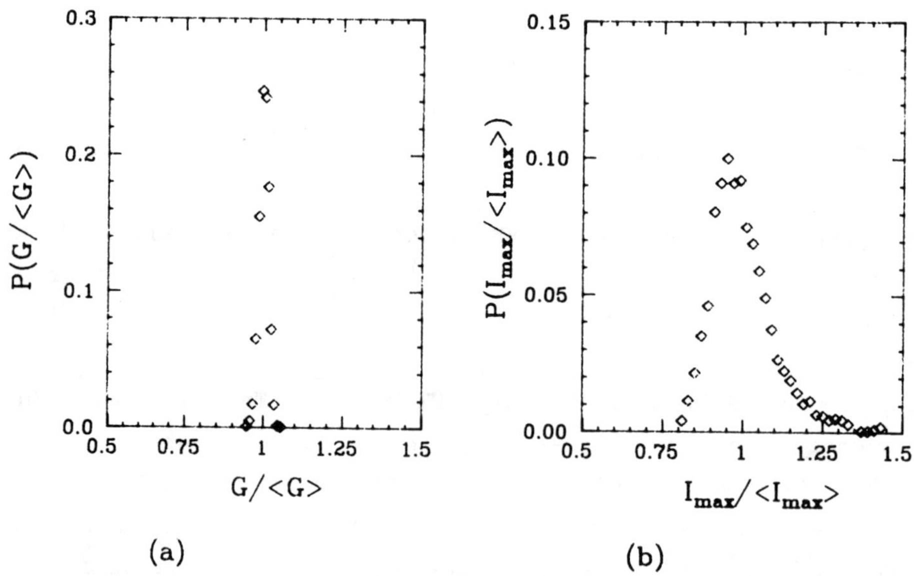

<u>Figure 5</u> The probability that a 50x50 RRN <u>a)</u> have conductance G and <u>b)</u> contain a maximum current I_{max} in any bond. These distributions were calculated from 2000 realizations of the RRN, and $\langle G\rangle$ and $\langle I_{max}\rangle$ are averages of the conductivity and maximum current respectively.

where we have calculated the conductivity and the size of the largest current occurring in the bonds of the network (for brittle systems this is proportional to the inverse of the electrical breakdown strength) in the network for a set of 2000, 50x50 random resistor networks at p=0.90. It is readily seen that the maximum current (or extreme moment) distribution is far broader than the conductivity (or low moment) distribution. The conductivity distribution is well represented by equation (37), while the breakdown distribution (Fig. 5b) is not. Two important features of extreme distributions (typified by the maximum current distribution of Fig. 5b), is that they are usually asymmetric, and they exhibit larger sample to sample fluctuations than low moment quantities such as moduli. It is possible to derive the form to be expected for the distribution given in Fig. 5b in several ways depending on the amount of mathematical rigor you wish to pursue. The most direct is to use the fact that we know that the distribution of bond currents is exponential for large currents for $p \gg p_c$. Then the cumulative probability, $C_1(I<I_{max})$, that a bond current is less than a maximum value, I_{max}, is given by,

$$C_1(I<I_{max}) \sim \int_0^{I_{max}} b\exp(-bI')dI' \tag{38}$$

$$= 1 - \exp(-bI_{max}) \tag{39}$$

If we consider the bond currents to be uncorrelated, then the probability that no current in the $N = L^d$ bonds in the network is greater than I_{max} is given by;

$$C_N(I<I_{max}) \sim (1 - \exp(-bI_{max}))^N \tag{40}$$

Since I_{max} is large, the exponential term is small, and the equation (40) is approximated by;

$$C_N(I<I_{max}) \sim \exp(-N\exp(-bI_{max})) \tag{41}$$

The probability that a system of size N has maximum current I_{max}, is then given by;

$$P(I^N_{max}) \sim bN \exp(-bI_{max}) \exp(-N \exp(-bI_{max})) \tag{42}$$

This form is clearly skew. The typical (or average) strength is found from,

$$N\exp(-b\langle I_{max}\rangle) \sim 1 \tag{43}$$

which implies

$$\langle I_{max}\rangle \sim \ln L \tag{44}$$

which has the correct finite size dependence leading to equation (29). A more complete analysis[5] that includes the dependence on dilution p shows that the cumulative distribution that is appropriate for the breakdown strength (in the case of electrical problems $I_b \sim 1/I_{max}$) is;

$$C_N(\sigma_b) = 1 - \exp(-kN \exp[-\xi_1 g_m^{1/\alpha}]) \text{ for mechanical systems} \tag{45}$$

where

$$\xi_1 = -\ln(1-p) \tag{46}$$

and

$$g_m = [(\sigma_0 - \sigma_b)/\sigma_b]/k_m \tag{47}$$

$$C_N(j_b) = 1 - \exp(-kN \exp[-\xi_1 g_e^{1/\alpha}]) \text{ for electrical systems} \tag{48}$$

where

$$g_e = [(j_0 - j_b)/j_b]/k_e \tag{49}$$

$$C_N(E_b) = 1 - \exp(-kN \exp[-\xi_1 g_d^{1/\alpha}]) \text{ for dielectric systems} \tag{50}$$

where

$$g_d = [(E_0 - E_b)/E_b]/k_d \tag{51}$$

The typical form of these cumulative distributions is shown in Fig. 6 for the electrical breakdown problem.

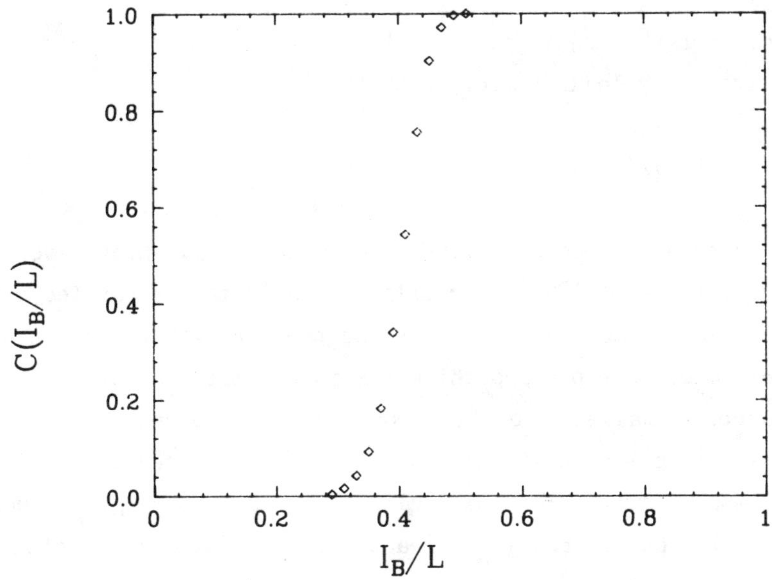

Figure 6 The probability that a network will fail when an external current of size I_b is applied. The distribution was calculated from 2000 configurations of 50x50 random resistor networks at p=0.90.

The expected sample to sample variations in strength are found by comparing the location of the 25% and 75% percentiles in these curves with the 50% percentile - the average value (these "percentiles" are more general concepts than the standard deviation and mean, which besides being most suitable for symmetric distributions, are especially bad indicators of observables when the local load distribution is very broad). Taking the electrical case as an illustrative example, we then find that the typical breakdown current value at the pth percentile is defined by;

$$c = p/100 = C(j_b^c) \tag{52}$$

Isolating j_b^c from this equation, we find that for $\ln N \gg \xi_1$,

$$j_b^c = j_0/(\ 1 + [\ln kN/\xi_1]^{\alpha}[\ 1 + A/\ln(kN)]) \tag{53}$$

where

$$A = -\alpha \ln(-\ln(1-c)) \tag{54}$$

The main point of this equation is that sample to sample variations in a system of size N are logarithmic, so that

$$\Delta j_b = O(K(p)/(\ln N)^2) \tag{55}$$

This is in contrast to conductivity or elastic moduli which have sample to sample variations $O(1/\sqrt{N})$. A similar analysis may be carried out for the dielectric and mechanical cases and the strength fluctuations are again logarithmic. For p approaching the pure limit, the strength fluctuations become smaller, so that exactly at the pure limit, they are zero ($K(p) \to 0$ as $p \to$ pure limit). A practical consequence of the large sample to sample fluctuations occurring in measured strength values, in combination with the destructive character of typical experiments, make experimental studies of breakdown phenomena more demanding than corresponding experiments of transport or elastic moduli.

In most engineering situations, the distribution used to fit breakdown data is of the form first proposed by Weibull[27], and in the mechanical case, the two parameter form of this distribution is,

$$C(\sigma) = 1 - \exp\left(-(\sigma/\sigma_0)^m\right) \tag{56}$$

where m is the Weibull modulus, or shape parameter, and σ_0 is the scale parameter. If the tail of the local load distribution is algebraic, carrying through arguments analogous to those leading to equations (38) - (41) leads to the form (56). In random resistor networks, such algebraic, or multifractal tails apply at the percolation point, so that it is expected that the Weibull form (56) should be appropriate there[28,29].

It is clear that near the percolation point, a new type of scaling analysis that takes into account the percolation correlation length, ξ,

$$\xi = (p-p_c)^{-\nu} \tag{57}$$

is necessary. In the next section we discuss the current state of our understanding of critical strength scaling and their attendant critical exponents.

IV STRENGTH SCALING NEAR PERCOLATION POINTS. Within the nodes links and blobs picture[30], as ξ becomes large (p approaching p_c), we may visualize the percolation backbone as a lattice of nodes with lattice spacing ξ, between which are links decorated by blobs. It is usually implicitly assumed that the node lattice is regular, and it is clear that this then predicts for electrical strength[4,5];

$$j_b \sim j_0 \xi^{-(d-1)} \sim j_0 (p-p_c)^{(d-1)\nu} \tag{58}$$

For dielectric problems, similar arguments lead to the result[4,5]

$$E_b \sim E_0/\xi \sim E_0 (p-p_c)^{\nu} \tag{59}$$

In mechanical problems, the situation is more complex, as the moment arm acting on a given link must be considered. An upper bound on the applied stress required to break the network is when this moment arm is assumed to be of order 1, so that

$$\sigma_b < \sigma_0 \, (p-p_c)^{(d-1)\nu} \tag{60}$$

The maximum size a moment arm can be of order the size of the distance between nodes so that,

$$\sigma_b > \sigma_0 \, (p-p_c)^{(d-1)\nu}/\xi \sim \sigma_0 \, (p-p_c)^{d\nu} \tag{61}$$

We thus believe that mechanical strength shows the critical behavior;

$$\sigma_b \sim \sigma_0 \, (p - p_c)^{x} \tag{62}$$

with;

$$(d-1)\nu < x < d\nu \tag{63}$$

These arguments apply equally well to central force networks and elastic models with bending forces. Bounds on x have also been calculated using energy balance arguments, and in this way, the following results have been derived[31,32];

$$f - \nu z < x < f-1 \tag{64}$$

$$(f + (d - d_b)\nu)/2 < x < (f + d\nu - 1)/2 \tag{65}$$

where, f is the elasticity exponent, z is the chemical distance exponent, and d_b is the fractal dimension of the percolation backbone. Although the exponent bounds above are different, none of them are inconsistent with experiments on perforated metal sheets[31,32]

From the fundamental equations (58) - (65), it is straightforward to include the conductivity factor to derive the breakdown electric field critical behavior of the fuse network[4,5]

$$E_b \sim (p-p_c)^{(d-1)\nu - t} \tag{66}$$

where t is the conductivity exponent. Similarly, the breakdown strain behavior of the elastic networks is given by;

$$\varepsilon_b \sim (p-p_c)^{x-f} \tag{67}$$

It is interesting to note that the breakdown electric field or strain may either diverge or go to zero near p_c depending on the relative size of the two exponents ν and t in equation (66) and x and f in equation (67). If E_b or ε_b diverge on approach to p_c, their behavior as a function of p is non-monotonic. That is they may decrease for small p due to dilute limit scaling (equations (29) to (31)) and diverge for p near p_c due to critical scaling (equations (66) and (67)). This non-monotonic behavior has been observed in several contexts[4,5,33].

It is possible to extend the critical scaling arguments discussed above to continuum percolation problems, where each link itself is composed of sublinks (necks) of various strength. The strength of a model then depends on the distribution of sublink strengths in addition to the long distance percolation geometry characterized by ξ. This sort of argument has been used to calculate the breakdown critical exponents relevant to electrical, dielectric and mechanical problems in "Swiss Cheese" and "Inverted Swiss Cheese" geometries[34,35].

For ξ large, but $L>\xi$, it is possible to combine the dilute limit scaling laws given in equations (29) and (31), with the critical scaling laws of equations (58) - (67), by replacing L in equations (29) and (30) by L/ξ. This substitution in equation (29) yields;

$$j_b \sim j_0 \xi^{-(d-1)} / [\, 1 + k_e (d \ln(L/\xi)/\xi_1)^{\alpha}] \tag{68}$$

This formula interpolates smoothly between the two singular behaviors in the limits $p \to$pure and $p \to p_c$, and provides a good approximation to strength behavior over the whole dilution range. A similar extension of equation (31) for the dielectric breakdown problem leads to,

$$E_b \sim E_0\, \xi^{-1} / [\, 1 + k_d (d \ln(L/\xi)/\xi_1)^{\beta}] \tag{69}$$

The mechanical fracture problem requires a slight modification of this procedure to yield,

$$\sigma_b \sim \sigma_0 \xi^{-x/\nu} / [\, 1 + k_m (d \ln(L/\xi)/\xi_1)^{\alpha}] \tag{70}$$

with x bounded as shown in equations (64) - (65).

V. <u>TOUGHENING OF RANDOM COMPOSITES</u>[16]. Mechanical strength is obviously a very important property in the engineering applications of materials. However, a property of equal importance in many applications is the <u>toughness</u> of the material. Toughness is a quantity that measures the amount of energy that a material can absorb before fracture, and in mechanical problems is related to the area under the stress strain curve. It is a good measure of impact resistance, flaw tolerance, and durability. Many metals are mechanically tough as a large amount of energy can be absorbed by dislocation emission before catastrophic fracture occurs. In metals, this toughening can be approximated by expanded zones near a crack tip, in which dislocation pile up occurs. With the development of high strength ceramic fibers that do not have significant toughening by dislocation emission, it has become necessary to develop novel toughening mechanisms, and many such mechanisms have been suggested. A good fraction of these mechanisms rely on stopping or deflecting a growing crack by the use of strong obstacles, and this sort of processes is especially relevant in composites such as fiber reinforced composites and polymers containing rubber inclusions. In such a mechanism, a crack may start growing, and then be stopped by an obstacle (e.g. a fiber). If the external stress is further increased, a crack may initiate in an entirely different part of the sample. This is a non-local toughening process which cannot be studied by single crack arguments alone, and has so far evaded a convincing analytical analysis. Some progress has been recently been made using percolation ideas and numerical modelling, and we now discuss those models and results.

As stated in Section II, the simplest percolation networks are brittle, as once a crack has begun to propagate at constant applied stress (mechanical breakdown), current (electrical breakdown) or electric field (dielectric breakdown), it usually leads to the failure of the whole network. Toughening of random networks has been studied using electrical and mechanical networks in two dimensions. Two mechanisms have been considered;

a) Toughening due to distributions of bond strengths[37].

This is essentially a crack deflection mechanism, as strong bonds lie in the path of any advancing crack, and cause it to change its propagation path locally. Numerical calculations and analytic arguments suggest that although this mechanism causes significant toughening (or ductility in the language of ref. 37) for smaller system sizes, a transition to brittle fracture appears to occur if the system is made large enough. These results were derived for a constant distribution of width $0 < w < 2$. Recent calculations of mechanical and electrical networks with Weibull distributions and algebraic distributions of bond strengths do not address this issue in detail, although some of the published stress-strain characteristics show evidence of toughening[22].

b) Toughening due to bonding behind the crack tip (residual bonding)[16].

As well as being of importance in the toughening of ceramics, this mechanism may occur in fiber-reinforced composites undergoing matrix fracture with fiber pullout. This is illustrated in Fig. 7, where the matrix crack is bridged by fibers that have undergone pullout, but which still provide a force holding the crack together.

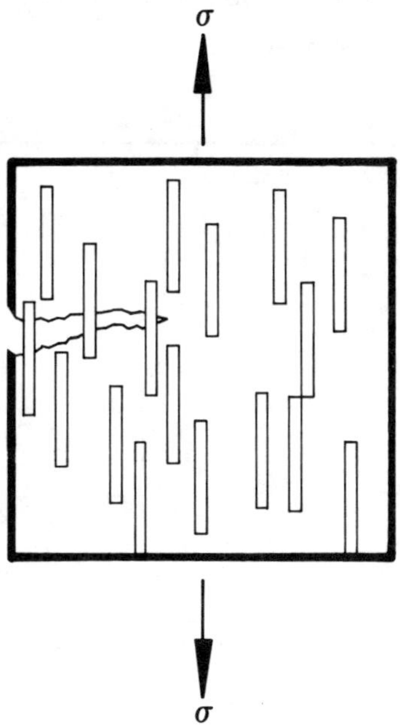

Figure 7 **Schematic example of a composite failing by fiber pullout. The unbroken fibers behind the crack tip reduce the stress intensity factor at the crack tip.**

We model this using networks, by assigning the bonds behind the crack tip a small elastic constant. The amount of residual bonding is then quantified by the ratio

$$R = \text{Residual Bonding Parameter} = e_0/e_r \quad (R_r/R_0) \tag{71}$$

where e_0 (R_0) is the background elastic constant (resistance), and e_r (R_r) is the elastic constant (resistance) of the residual bonds. The effect of these residual bonds on the stress intensity at the tip of the crack is shown in Fig. 8. In this calculation, a crack (void) of initial size 6 (or 8) atomic spacings was placed in a 40x40 central force network, and an external tensile stress applied perpendicular to the long axis of the crack. Bonds change from e_0 to e_r when a critical stress occurs across them. It is seen from Fig.8, that after an initial increase in stress intensity, the residual bonding behind the crack tip carries enough of the applied load to actually reduce the stress intensity at the crack tip. At a fixed applied stress that is sufficient to initiate crack growth, is thus not sufficient to cause catastrophic failure, and this crack shows crack arrest.

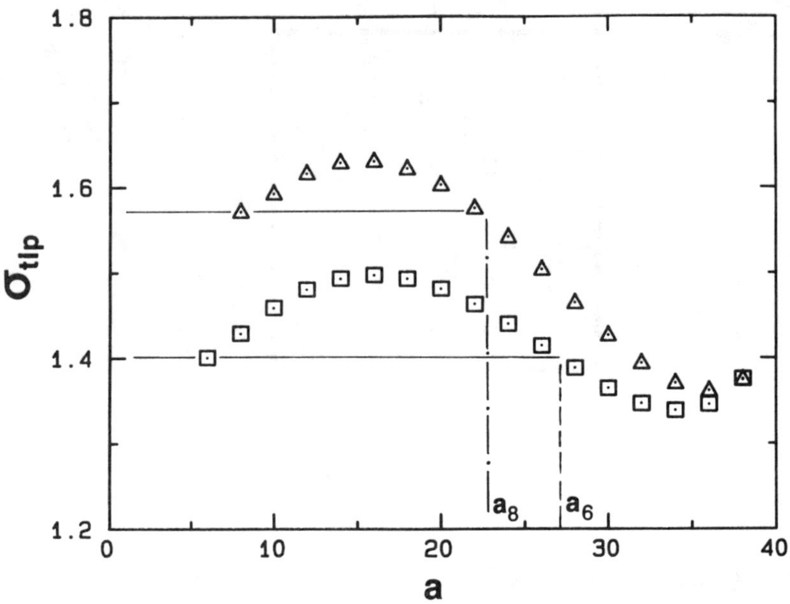

Figure 8 The stress intensity factor at the tip of a crack which has residual bonding (R=20) behind the crack tip. (□) an initial crack of size 6, falls below its initial stress intensity factor at length a_6, (Δ) an initial crack of size 8 falls below its initial stress intensity factor at length a_8. When the external stress is just large enough to initial crack growth, crack arrest occurs at lengths a_6 and a_8, respectively.

A similar effect may be produced in electrical networks, where the analogy of the residual elasticity is a (small) residual conductivity that remains after a bond has failed electrically. This electrical problem is more tractable numerically than the mechanical networks, and provides a good pedagogical model in which to study this toughening mechanism.

If a large amount of crack arrest occurs, many local regions suffer some damage before the whole system fails. The brittle I-V characteristics of the simplest percolation networks are changed by this process, as shown in Fig. 9 for random resistor networks at p=0.75, with different residual bonding parameters R.

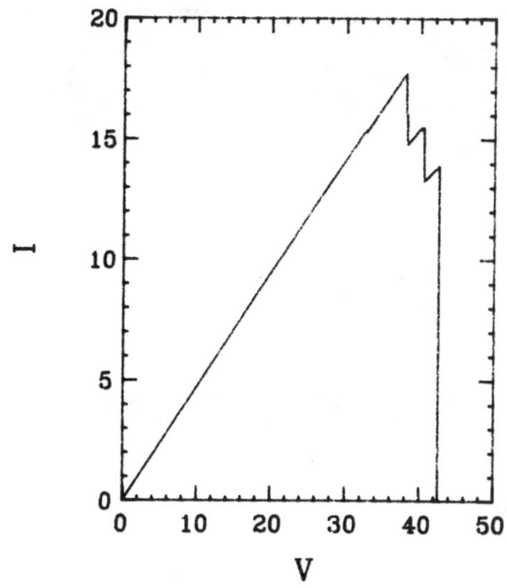

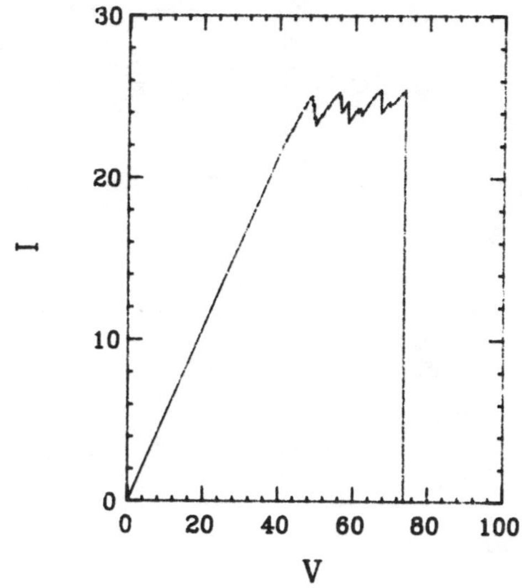

Figure 9 I-V characteristics for 70x70 resistor networks with residual bonding a) R=100; and b) R=20.

It is clear from this figure, that at R=20, the failure of the first bond and the failure of the whole network occur at different loads, and that the network is significantly toughened. A quantitative measure of toughening in these problems, is the number of bonds that fail before the whole system fails. For brittle problems, we expect

$$N_b^b \sim L^{d-1} \text{ or } n_b = N_b/L^{d-1} \sim \text{constant for brittle fracture} \tag{72}$$

as a cleavage crack cuts a surface, while for highly toughened networks, we expect,

$$N_b^t \sim L^d \text{ or } n_b = N_b/L^{d-1} \sim L^y \quad \text{with } y>0 \tag{73}$$

y is the toughening exponent, and if y>0, it indicates that damage occurs in many parts of the system, and hence that the material toughens by a non-local mechanism. n_b is then an "order parameter" for the transition from brittle to tough networks. A plot of n_b against R is shown in Fig. 10 for p=0.75.

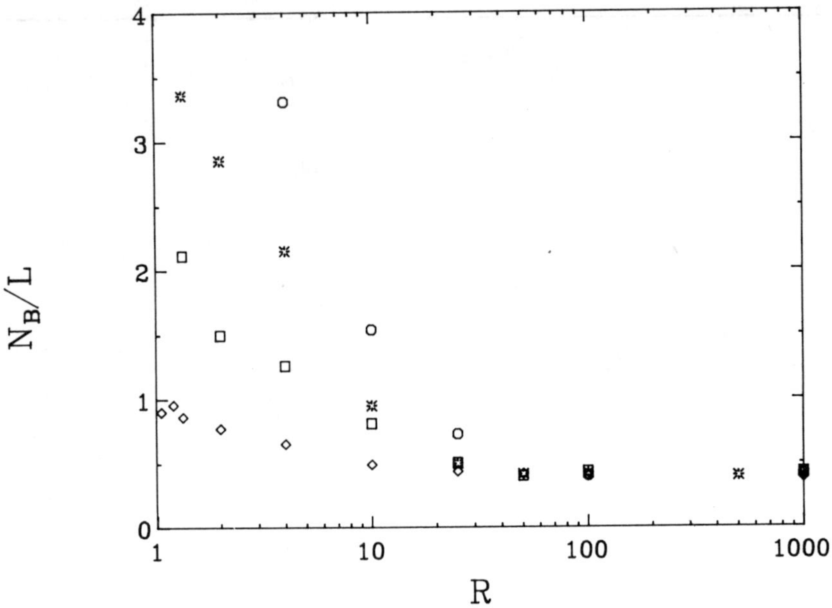

Figure 10 **The number of bonds broken in the network failure as a function of the residual bonding parameter R. L=10 (<>); L=20 (□); L=30 (*); L=40 (o).**

As $R \to \infty$, equation (72) holds and the network is brittle. For R small however, n_b increases strongly with system size, indicating a toughening trend, and $y > 0$ as indicated in equation (73). An interesting feature of Fig. 10 is that there appears to be a sharp transition between the brittle and tough regimes at $R_c \sim 35$, so that for $R > R_c$ the network is brittle and that for $R < R_c$ it is tough. If this sort of result extends to mechanical systems, the existence of such critical toughening parameters is of considerable practical as well as its intrinsic intellectual interest. We close this section by noting that a reliable analytic approximation for the onset of such toughening transitions either in the residual bonding networks or in networks with bonds having continuous distributions of strengths, is still lacking.

VI. **OVERVIEW AND FUTURE PERSPECTIVE**. Having reviewed work on networks models of breakdown, we now place these calculations in a broader context, by pointing to related work going on in the materials science, engineering, geophysics and statistical mechanics communities. This overview does not exhaustively list papers in these areas, but rather lists papers that give a good introduction to the methods, ideas and problems of major interest at the present time.

MODELS FOR FIBER BUNDLES AND FIBER REINFORCED COMPOSITES

In the network models discussed in this paper, the computationally intensive part of the crack propagation process is the redistribution of load after local bond failure. In contrast, a large amount of work by researchers in the fiber and the fiber reinforced composite communities has been devoted to fiber bundle models[38-45] with equal or local load sharing rules. In these systems the load that was on a failed bond (or fiber unit) is shared with a specified number of neighbors. If all parallel neighbors share the load equally, the model is called equal load sharing[38-40] (this is more appropriate to loosely wound fibers), while if only a few local neighbors share the load, the model is called local load sharing[41-44] (this is more appropriate to well bonded fibers in fiber reinforced composites). These "ad-hoc" load sharing rules greatly reduce the computational effort, but contain several approximations. A detailed comparison of the load sharing rule models with the models which use linear relaxation algorithms to redistribute load, for a given microstructure, could lead to a greater understanding of both classes of model.

MODELS FOR POROUS MATERIALS AND EARTH-QUAKE MODELS

Modelling of static strength of porous materials was, till recently, largely empirical. Kendall[48] reviews work till 1984, and discusses various laws that incorporate an exponential decrease in mechanical strength as a function of pore fraction with the "Griffith" square root reduction in strength that occurs due to a large crack. Statistical mechanics and renormalization group ideas have been applied to earthquake propagation and statistics using static and dynamic models[49-56]. Some of these models are very similar to the fiber bundle models of fibrous composites[38-45] discussed above. The concept of fracture in a fractal environment is of intrinsic interest[52-60], and has been used in earthquake models[52-56]. There have also been recent attempts to correlate the fractal dimensions of fracture surfaces to toughness in metals[61,62], although this correlation is still controversial.

NON-EQUILIBRIUM GROWTH MODELS OF CRACK GROWTH AND BREAKDOWN

This subject was initiated by the observation that certain diffusion limited aggregation (DLA) models[63-67] can be interpreted as models for dielectric breakdown. These Laplacian models have been generalized to linear elastic problems[67-72], and there, the fractal dimension of the dynamically grown cracks is the major focus of the research. The fiber bundle models discussed above have also been generalized to the dynamic case, and studied extensively[45-47]. There the time to failure is the quantity of major interest.

CONNECTION WITH R-CURVE BEHAVIOR

Many materials exhibit an increasing resistance to fracture, increased toughness, with crack extension[73]. In ceramics, one important theory for the origin of this toughening mechanism is that there is some ligamentary bonding behind the crack tip[74] This observation is what motivated the calculations of the residual bonding model described in Section V, and there is considerable scope for the application of network models to this and other toughening mechanisms in brittle materials.

We close this overview section by noting that the number of properties labelled as "breakdown phenomena" continues to broaden, and that the ideas developed to study extreme fluctuations in breakdown problems, may eventually impact on many other statistical mechanics and disordered system problems which are dominated by extreme fluctuations (e.g. Lifshitz tails in semiconductor band gaps, Griffith singularities, relaxation rates in disordered systems, and nucleation and growth)

REFERENCES

1. H. Takayasu, Simulation of Electric Breakdown and Resulting Variant of Percolation Fractals, Phys. Rev. Letts. 54, 1099 (1985).

2. L. de Arcangelis, S. Redner and H.J. Herrmann, A Random Fuse Model for Breaking Processes, J. Physique Letts. 46, L585 (1985).

3. M. Sahimi and J.D. Goddard, Elastic Percolation Models for Cohesive Failure in Heterogeneous Systems, Phys. Rev. B33, 7848 (1986).

4. P.M.Duxbury, P.D. Beale and P.L. Leath, Size Effects of Electrical Breakdown in Quenched Random Media, Phys. Rev. Letts. 57, 1052 (1986).

5. P.M. Duxbury, P.L. Leath and P.D. Beale, Breakdown Properties of Quenched Random Systems: The Random-Fuse Network, Phys. Rev. B36, 367, (1987).
Y.S. Li and P.M. Duxbury, Size and Location of the Largest Current in a Random Resistor Network, Phys. Rev. B36, 5411 (1987).

6. P.D. Beale and P.M. Duxbury, Theory of Dielectric Breakdown in Metal-Loaded Dielectrics, Phys. Rev. B37, 2785 (1988).

7. P.D. Beale and D.J. Srolovitz, Elastic Fracture in Random Materials, Phys. Rev. B37, 5500 (1988).

8. R.W. Davidge, Mechanical Behavior of Ceramics, Cambridge University Press (1979).

9. J.C. Garland and D.B. Tanner eds., Electrical Transport and Optical Properties of Inhomogeneous Media, AIP Conf. Proc 40 (1978).

10. G. Deutscher, R. Zallen and J. Adler eds., Percolation Structures and Processes, Ann. Israel Phys. Soc. 5 (1983).

11. Y.S. Li and P.M. Duxbury, From Moduli Scaling to Breakdown Scaling-A Moment Spectrum Analysis, Preprint.

12. D.J. Bergman, Non-Linear Behavior and 1/f Noise Near a Conductivity Threshold - The Effects of Local Microgeometry, Phys. Rev. B, Preprint. - see also Dr. Bergman's contribution to these proceedings.

13. E.J. Gumbel, Statistics of Extremes, Columbia Univ. Press, New York (1958).

14. E. Castillo, Extreme Value Theory in Engineering, Academic Press (1988).

15. P.M. Duxbury and P.L. Leath, The Failure Distribution in Percolation Models of Breakdown, J. Phys. A20, L411 (1987).

16. Y.S. Li and P.M. Duxbury, Crack Arrest by Residual Bonding in Resistor and Spring Networks, Phys. Rev. B38, 9257 (1988).

17. see e.g. Chap. 3 of Ref. 5.

18. E.J. Garboczi, Linear Dielectric-Breakdown Electrostatics, Phys. Rev. B38, 9005 (1988). Note that the analogy between electrical and mechanical breakdown drawn in this paper is only valid in two dimensions.

19. D.R. Bowman and D. Stroud, Theory of Dielectric Breakdown in Metal-Loaded Composites, (Unpublished).

20. M. Soderburg, Resistive Breakdown of Inhomogeneous Media, Phys. Rev. B35, 352 (1987).

21. A. Gilabert, C. Vanneste, D. Sornette and E. Guyon, The Random Fuse Network as a Model of Rupture in a Disordered Medium, J. Physique 48, 763 (1987).

22. H.J.Herrmann, A. Hansen and S. Roux, Fracture of Disordered, Elastic Lattices in Two Dimensions, Phys. Rev. B39, 637 (1989).

23. G.N. Hassold and D.J. Srolovitz, Brittle Fracture in Randomly Defected Materials, Preprint.

24. R.W. Coppard, L.A. Dissado, S.M. Rowland and R. Rakowski, Dielectric Breakdown in Metal Loaded Polyethylene, Preprint.

25. C. Laurent, E. Kay and N. Souag, Dielectric Breakdown of Polymer Films Containing Metal Clusters, J. Appl. Phys. June 1988.

26. R. M. Fulrath and J. A. Pask eds., Ceramic Microstructures, New York (Wiley 1966), see page 367.

27. W. Weibull, A Statistical Theory of the Strength of Materials Proc. Ing. Vetenskapsakad No. 151 (in English).

28. M.D. Stephens and M. Sahimi, Distribution of Fracture Strengths in Disordered Continua, Phys. Rev. B36, 8656 (1987).

29. D. Sornette, Weibull-Like Failure Distribution Induced by Fluctuations in Percolation, J. Phys. France 49, 889 (1988).

30. see e.g. D. Stauffer, Introduction to Percolation Theory, Taylor and Francis (London, 1985).

31. L Benguigui, P. Ron and D.J. Bergman, Strain and Stress at the Fracture of Percolative Media, J. Physique 48, 1547 (1987).

32. B.K. Chakrabarti, Mechanical and Electrical Breakdown in Disordered Solids, Reviews of Solid State Science 2, 559 (1988).

33. L. de Arcangelis, S. Redner and A. Coniglio, Multiscaling Approach in Random Resistor and Random Superconducting Networks, Phys. Rev. B34, 4656 (1986) - Fig. 10.

34. B.K. Chakrabarti, A.K. Roy and S.S. Manna, Breakdown Exponents in Lattice and Continuum Percolation, J. Phys. C21, L65 (1988).

35. C.J. Lobb, P.M. Hui and D. Stroud, Nonuniversal Breakdown Behavior in Superconducting and Dielectric Composites, Phys. Rev. B36, 1956 (1987).

36. D. Sornette, Critical Transport and Failure in Continuum Crack Percolation, J. Phys. France 49, 1365 (1988).

37. B. Kahng, G.G. Bartrouni, S. Redner, L. De Arcangelis and H.J. Herrmann, Electrical Breakdown in a Fuse Network with Random, Continuously Distributed Breaking Strengths, Phys. Rev. B37, 7625 (1988).

38. H.E. Daniels, The Statistical Theory of the Strength of Bundles of Threads, Proc. Roy. Soc. 183A, 405 (1945).

39. B.D. Coleman, On the Strength of Classical Fibers and Fiber Bundles, J. Mech. and Phys. of Solids, 7, 60 (1958).

40. R.L. Smith and S.L. Pheonix, Asymptotic Distributions for the Failure of Fibrous Materials Under Series-Parallel Structure and Equal Load Sharing, J. Appl. Mech. 48, 75 (1981).

41. D.G. Harlow and S.L. Pheonix, The Chain-of-Bundles Probability Model For the Strength of Fibrous Materials I: Analysis and Conjectures, J. Comp. Mats. 12, 195 (1978).

42. R.L. Smith, S.L. Pheonix, M.R. Greenfield, R.B. Henstenburg and R.E. Pitt, Lower-Tail Approximations for the Probability of Failure of Three-Dimensional Fibrous Composites with Hexagonal Geometry, Proc. Roy. Soc. A388, 353 (1983).

43. R.L. Smith and H.W. Taylor, Models for Fiber-Matrix Composites with Local Load Sharing, Operations. Research, 32, 649 (1984).

44. S.B. Batdorf and R. Ghaffarian, Size Effect and Strength Variability of Unidirectional CompositesInt. J. Fract. 26, 113 (1984).

45. B.D. Coleman, Statistics and Time Dependence of Mechanical Breakdown in Fibers, J. Appl. Phys. 29, 968 (1958).

46. S.L. Pheonix and L.J. Tierney, A Statistical Model for the Time Dependent Failure of Unidirectional Composite Materials Under Local Elastic Load-Sharing Among Fibers, Eng. Fract. Mech. 18, 193 (1983).

47. C.C. Kuo and S.L. Pheonix, Recursions and Limit Theorems for the Strength and Lifetime Distributions of a Fibrous Composite, J. Appl. Prob. 24, 123 (1987).

48. K. Kendall, Connection Between Structure and Strength of Porous Solids, AIP Conf. Proc. 107, 78 (1984).

49. C.J. Allegre, J.L. Le Mouel and A. Provost, Scaling Rules in Rock Fracture and Possible Implications for Earth-Quake Prediction, Nature 297, 47 (1982).

50. W.I. Newman and L. Knopoff, A Model for Repetitive Cycles of Large Earthquakes, Geophys. Res. Letts. 9, 735 (1982).

51. W.I. Newman and L. Knopoff, Crack Fusion Dynamics: A Model for Large Earthquakes, Geophys. Res. Letts. 10, 305 (1983).

52. T.L. Cheldize, Percolation Theory as a Tool for Imitation of Fracture Process in Rocks, Pageoph. 124, 731 (1986).

53. D.L. Turcotte, R.F. Smalley Jr and S.A. Solla, Collapse of Loaded Fractal Trees, Nature 313, 671 (1985).

54. R.F. Smalley Jr, D.L. Turcotte and S.A. Solla, A Renormalization Group Approach to the Stick-Slip Behavior of FaultsJ. Geophys. Res. 90, 1894 (1985).

55. J. Lomnitz-Adler, Asperity Models and Characteristic Earthquakes Tectonophys. 120, 133 (1985).

56. J. Lomnitz-Adler, On the Magnitude-Frequency Relation of Asperity Models, Geophys. J. R. Atron. Soc. 83, 435 (1985).

57. R.B. Stinchcombe, P.M. Duxbury and P. Shukla, The Minimum Gap on Diluted Cayley Trees, J. Phys. A19, 3903 (1986).

58. D. Sornette, Elasticity and Failure of a set of Elements Loaded in Parallel, Preprint.

59. W.I. Newman and A.M. Gabrielov, Failure of Heirarchical Distributions of Fiber Bundles I, Preprint.

60. R.F. Cook, Effective-Medium Theory for Fracture of Fractal Porous Media, Preprint.

61. B. Mandelbrot, D.E. Passoja and A.J. Paullay, Fractal Character of Fracture Surfaces of Metals, Nature 308, 721 (1984).

62. E.E. Underwood and K. Banerji, Fractals in Fractography Mat. Sci and Eng. 80, 1 (1986).

63. L. Niemeyer, L. Pientronero and H.J. Wiesmann, Fractal Dimension of Dielectric Breakdown, Phys. Rev. Letts. 52, 1033 (1984).

64. M. Murat, Dielectric Breakdown Between Parallel Plates Phys. Rev. B32, 8420 (1985).

65. H.J. Wiesmann and H.R. Zeller, A Fractal Model of Dielectric Breakdown and Prebreakdown in Solid Dielectrics, J. Appl. Phys. 60, 1770 (1986).

66. L. Pietronero, A. Erzan and C. Evertsz, Theory of Laplacian Fractals: Diffusion Limited Aggregation and Dielectric Breakdown Model, Physica A151, 207 (1988).

67. H.E. Stanley and N. Ostrowsky eds, Random Fluctuations and Pattern Growth - Experiments and Growth, Kluwer Academic (1988).

68. Y. Termonia and P. Meakin, Formation of Fractal Cracks in a Kinetic Fracture Model, Nature 320, 429 (1986).

69. E. Louis and F. Guinea, The Fractal Nature of Fracture, Europhys. Letts. 3, 871 (1987).

70. L. Fernandez, F. Guinea and E. Louis, Random and Dendritic Patterns in Crack Propogation, J. Phys. A21, L301 (1988).

71. A.T. Skjeltorp and P. Meakin, Fracture in Microsphere Monolayers Studied by Experiment and Computer Simulation Nature 335, 424 (1988).

72. P. Meakin, G. Li, L.M. Sander, E. Louis and F. Guinea, A Simple Two Dimensional Model for Crack Propogation, Preprint.

73. R.F. Cook and D.R. Clarke, Fracture Stability, R-Curves and Strength Variability, Acta Metall. 36, 555 (1988).

74. R.F. Cook, C.J. Fairbanks, B.R. Lawn and Y.W. Mai., Crack Resistance by Interfacial Bridging: Its Role in Determining Strength Characteristics, J. Mater. Sci. 2, 345 (1987).

CHAPTER 9

Convexity in Random Resistor Networks*

Kenneth Golden†

Abstract. The bulk conductivity $\sigma^*(p)$ of the bond lattice in $\mathbb{Z}^d$ is considered, where the conductivity of the bonds is either 1 with probability p or $\epsilon \geq 0$ with probability $1-p$. Rigorous and non-rigorous results demonstrating convexity of $\sigma^*(p)$ near the percolation threshold p_c are presented. For $\epsilon = 0$, a certain transformation on graphs which drives the system to p_c is found to be "convexity improving". This analysis leads us to propose upper and lower bounds on the conductivity critical exponent t in terms of some percolation exponents. These bounds become tighter with increasing dimension and coincide in $d = 6$, where mean field behavior is believed to hold.

1. Introduction. Disordered conductors are encountered frequently in science and engineering. Examples include inhomogeneous gold films on glass, impure semiconductors, suspensions of particles in a fluid, composite materials, and even sea ice. Of particular practical and theoretical interest are those systems which undergo an insulator/conductor transition as some parameter is varied. For example, sea ice is composed of salty water, or brine inclusions (conductor) embedded in a pure ice matrix (insulator). When the sea ice is cold, the brine pockets occupy only a small volume fraction and the sea ice behaves as an insulator. However, when the sea ice is warmer, the brine pockets tend to coalesce and form a conducting matrix, so that the sea ice behaves as a conductor.

In the late 1960's Ziman [**1**], Eggarter and Cohen [**2, 3**] and others suggested that random resistor networks based on the percolation model [**4**] provide a good

*Supported in part by NSF and AFOSR through NSF Grant DMS - 8801673.

† Department of Mathematics, Princeton University, Princeton, NJ 08544 USA

description of disordered conductors. In particular, consider the bulk conductivity $\sigma^*(p)$ of the bond lattice in $\mathbb{Z}^d$, where the conductivity of the bonds is either 1 with probability p, or $\epsilon \geq 0$ with probability $1-p$. When $\epsilon = 0$, we can view the conductivity 0 bonds as vacant. In this case, the associated percolation problem concerns $P_\infty(p)$, the probability that the origin is connected to an infinite cluster of occupied (conductivity 1) bonds. For p below some critical probability p_c, called the percolation threshold, $P_\infty(p) = 0$, while for $p > p_c$, $P_\infty(p) > 0$. The bulk conductivity $\sigma^*(p)$ has a similar behavior, with $\sigma^*(p) = 0$ for $p < p_c$, and $\sigma^*(p) > 0$ for $p > p_c$, although there is apparently no simple relation between $\sigma^*(p)$ and $P_\infty(p)$ [**5**]. As $p \to p_c^+$, it is believed that $\sigma^*(p)$ exhibits critical scaling, $\sigma^*(p) \sim (p-p_c)^t$, where t is called the conductivity critical exponent [**6**].

Since their introduction, random resistor networks have been widely studied in the physics literature [**7, 8**]. Given their central place in the theory of disordered conductors, it is surprising that there has been little rigorous analysis of random resistor networks. The main exceptions are the works of Grimmett and Kesten [**9**] (see also ref. [**10**]) and Chayes and Chayes [**11, 12**]. One of the principal contributions of these works is to establish the coincidence of the conduction and percolation thresholds for $d = 2$, where $p_c = \frac{1}{2}$ [**9**], as well as for higher dimensions [**11**]. In addition, the Chayes [**11**] obtain bounds on t (assuming it exists) in terms of some percolation exponents. In $d = 2$ it has also been established, using arguments that can be made rigorous, that $t \geq 1$ [**13, 14,** see also ref. **12**]. Furthermore, it is rigorously known that $\sigma^*(p)$ is continuous at $p_c = \frac{1}{2}$ in $d = 2$, which is proven via the continuity of $P_\infty(p)$ at $p_c = \frac{1}{2}$ [**15**] and the bound $\sigma^*(p) \leq pdP_\infty^2(p)$ [**11**]. In higher dimensions it is certainly believed that $\sigma^*(p)$ is continuous at p_c, but this has not been rigorously proven yet.

The purpose of the present work is to introduce a new approach to the random resistor problem based on convexity, which has played an important role in many problems of statistical physics, but has apparently all but been ignored in the present context. As Straley [**16**] remarks, $\sigma^*(p)$ is " necessarily positive but has no convexity property, with the consequence that no rigorous exponent inequalities can be proved." Nevertheless, casual inspection of numerical simulations [**6, 17-20**] of the graph of $\sigma^*(p)$ for bond or site models in $d \geq 2$ suggests convexity in p, at least near the percolation threshold p_c. (In the site problem, vertices of $\mathbb{Z}^d$, along with all $2d$ attached bonds, are removed (when $\epsilon = 0$) at random with probability $1-p$.) Given the broad and enduring interest in these models, we believe that it is important to investigate this convexity, which appears to be a general feature of the conductivity of lattices near p_c.

The principal results discussed in this paper are as follows. First we observe directly that $\sigma^*(p)$ cannot be convex for all p when $\epsilon = 0$, and present numerical results outlining the regimes of ϵ and p for which $\sigma^*(p)$ is convex. Next, rigorous results for $\epsilon > 0$ are obtained, the main one being that $\sigma^*(p)$ for the $d = 2$ bond problem is convex near $p_c = \frac{1}{2}$ for every $\epsilon > 0$. The proof is based on Keller's Interchange Theorem, which holds for certain continuum systems, so that our result holds for them as well. Finally the $\epsilon = 0$ case is considered, and a general physical argument explaining convexity near p_c is offered. Our physical argument is supported by a rigorous result asserting that a certain transformation which maps the $p = 1$ lattice to a well-known model of the conducting backbone near p_c, namely the node-link model, is "convexity improving". Further analysis leads us to propose upper and lower bounds on t, in terms of some percolation exponents, which become tighter with increasing dimension, and coincide for $d = 6$, where mean field behavior is believed to hold, with $t = 3$.

2. Formulation. We formulate the bond conductivity problem for an arbitrary graph. Let G be a finite graph consisting of N bonds $\{b_i\}$ and N' vertices $\{x_i\}$. Assigned to G are N independent random variables c_i, $1 \leq i \leq N$, the bond conductivities, which take the values 1 with probability p and $\epsilon \geq 0$ with probability $q = 1 - p$. Distinguish two vertices, say $x_1 = x$ and $x_{N'} = y$, and connect them to a battery which keeps the voltage drop between them equal to 1. The effective conductivity $\tilde{\sigma}(\omega)$ of the network for any realization ω of the bond conductivities is just the total current $i(\omega)$ that flows through the network, which is obtained via Kirchoff's laws. We define $\sigma(p) = <\tilde{\sigma}(\omega)>$, where the expectation $<\cdot>$ is over all 2^N realizations. For example, a two bond network has

$$\sigma(p) = p^2\tilde{\sigma}(1,1) + pq\big(\tilde{\sigma}(1,\epsilon) + \tilde{\sigma}(\epsilon,1)\big) + q^2\tilde{\sigma}(\epsilon,\epsilon) \ ,$$

where $\tilde{\sigma}(1,1) = \tilde{\sigma}(\omega)$ with $\omega = (1,1)$, and so on. For N bonds, $\sigma(p)$ is an Nth order homogeneous polynomial in p and q,

$$\begin{aligned} \sigma(p) &= \sum_{k=0}^{N} \alpha_k p^{N-k} q^k \\ \alpha_k &= \sum_{\omega^k \in \Omega^k} \tilde{\sigma}(\omega^k), \quad q = 1 - p, \end{aligned} \tag{2.1}$$

where $\Omega^k = \big\{\omega^k = (\omega_1, \ldots, \omega_N) | \omega_\ell = \epsilon \text{ for exactly k of the } \omega_\ell\text{'s}\big\}$.

The cases of most interest are when G is a square, cubic, or hypercubic lattice. Then, with $d = 2$ for simplicity, we take an $L \times L$ sample of the lattice and attach a perfectly conducting bus bar to each of two opposite edges of the sample. This can be accomplished [**9**] in the above language by attaching to each vertex of these opposing edges a perfectly conducting bond. All of these bonds from one edge meet at a new vertex x and all the bonds from the other edge meet at a new vertex y. Then x and y are connected again with the unit battery. Random bond conductivities are assigned only to the bonds in the original $L \times L$ sample. Let $\sigma_L(p)$ denote (2.1) for the effective conductivity measured between x and y. Then for $d \geq 1$, the bulk conductivity $\sigma_L^*(p)$ is defined as

$$\sigma_L^*(p) = L^{2-d}\sigma_L(p) \ . \tag{2.2}$$

For $\epsilon > 0$, Künnemann [**21**] proved that the infinite volume limit

$$\sigma^*(p) = \lim_{L\to\infty} \sigma_L^*(p) \tag{2.3}$$

exists. For $\epsilon = 0$, the existence of (2.3) has still not been proven [**9-11**]. In what follows we shall assume that this limit exists. It should be remarked that since for $\epsilon > 0$, $\sigma^*(p,\epsilon) = \lim_{L\to\infty} \sigma_L^*(p,\epsilon)$ exists and is monotonically decreasing in ϵ, $\lim_{\epsilon\to 0} \sigma^*(p,\epsilon)$ exists, and provides a reasonable definition for the bulk conductivity in the $\epsilon = 0$ case. The unsolved problem of the existence of (2.3) directly for $\epsilon = 0$ then boils down to whether or not the $\epsilon \to 0$ and $L \to \infty$ limits can be interchanged.

3. Regimes of Convexity. We now make an observation which directly indicates that $\sigma^*(p)$ cannot generally be a convex function of p. For the $d = 2$ bond problem with $\epsilon = 0$, it is known [**7, 19**] (although not rigorously) that $\frac{d\sigma^*}{dp}|_{p=1} = 2$. Now, a straight line with slope 2 at $\sigma^*(1) = 1$ intersects the p-axis at $p = \frac{1}{2}$. If $\sigma^*(p)$ is convex for all p, then either $p_c < \frac{1}{2}$ or the graph of $\sigma^*(p)$ is the above straight line, which is the effective medium solution for this problem. Since it is rigorously known [**9**] that $p_c = \frac{1}{2}$ for $d = 2$, the only way $\sigma^*(p)$ could be convex for all p is if effective medium theory gives the correct solution for all p, which gives a critical exponent of $t = 1$ and contradicts practically every numerical simulation of this problem. From inspection of the simulations in references [**6**] or [**19**], as p increases from $\frac{1}{2}$ to 1, what apparently happens is that $\sigma^*(p)$ starts off convex at $p = \frac{1}{2}$, but eventually the curve "turns over", i.e. becomes slightly concave, which allows it to have the correct slope of 2 at $p = 1$. We note that this effect is subtle, because away from the critical regime ($p \approx \frac{1}{2}$), the graph of $\sigma^*(p)$ looks nearly linear, where effective medium theory is believed to provide a very good approximation. In fact, an expression for $\frac{d^2\sigma^*}{dp^2}|_{p=1}$ is found in reference [**19**] and is numerically evaluated, with a result of about $-.21$, which supports the accuracy of effective medium theory near $p = 1$. The calculations of $\frac{d^2\sigma^*}{dp^2}|_{p=1}$ are somewhat involved, and it is not at all obvious before numerical evaluation that the result will be negative, hence the need for our above direct argument.

Another direct way of seeing that convexity is not a general principle is the following finite network pointed out to the author by P. Doyle. Let the network consist of three circuit elements in series, the first consists of two bonds in parallel, the second (and middle) is a single bond, and the third is again two bonds in parallel. When $\epsilon = 0$, an elementary calculation shows $\frac{d^2\sigma}{dp^2}|_{p=1} < 0$.

In order to build intuition about convexity in resistor networks, we have examined $\frac{d^2\sigma}{dp^2}$ for a variety of finite graphs. The calculations are based on an exact formula [**22, 23**]for the effective conductivity $\widetilde{\sigma}(\omega)$ of a conductor network based on a graph

G for any realization of the bond conductivities c_i,

$$\widetilde{\sigma}(\omega) = \frac{\sum_{T} \prod_{b_i \in T} c_i}{\sum_{T_{xy}} \prod_{b_i \in T_{xy}} c_i} \quad , \tag{3.1}$$

where the sum in the numerator is over all spanning trees T in G, and the sum in the denominator is over all spanning trees T_{xy} in G with the vertices x and y identified. The numerator and denominator are computed via the determinants of the appropriate adjacency matrices. The graphs we have considered include the square lattice, triangular lattice, trees (Bethe lattice), ladders, Wheatstone bridge, and others. We have considered bond and site problems, and various ranges of $\epsilon \geq 0$. Most of the networks we have looked at have been rather small, the largest being a 15×15 sample of the square lattice.

To summarize our results, away from $p = 1$ all networks we have considered have $\sigma(p)$ convex for any $\epsilon \geq 0$, and for both bond and site problems. Near $p = 1$, however, bond problems typically are not convex when ϵ is small enough, while site problems typically are convex for all $\epsilon \geq 0$. For a typical bond problem, if we start with ϵ close to 1 and allow it to shrink, $\frac{d^2\sigma}{dp^2}|_{p=1}$ is positive until, say $\epsilon \approx 0.1$, below which it becomes and stays negative all the way down to $\epsilon = 0$. This concavity near $p = 1$ for $\epsilon = 0$ is, of course, consistent with the results in [**19**]. Furthermore, the convexity near $p = 1$ for site problems is consistent with the results of [**20**]. We see then for the site problem, $\sigma^*(p)$ for the square lattice with $\epsilon = 0$ appears to be convex for all p. If indeed this is the case, then one can obtain a bound on p_c using the result [**17, 20**] for $d = 2$ that $\frac{d\sigma^*}{dp}|_{p=1} = \pi$. The bound would be obtained by drawing a straight line of slope π through the point $p = 1$, $\sigma^*(p) = 1$, and noting that p_c is less than the intercept of this line, which leads to

$$p_c < 1 - \frac{1}{\pi} \; . \tag{3.2}$$

For the $d = 2$ site problem, p_c is believed [**17**] to be about .59, whereas $1 - 1/\pi \approx .68$.

4. Convexity when $\epsilon > 0$. In this section we directly consider the infinite volume limit $\sigma^*(p)$ for the bond problem in dimensions $d \geq 2$ with $\epsilon > 0$. In order to prove our principal results, we shall need some smoothness in p of $\sigma^*(p)$. The analysis for this is based on an integral representation for $\sigma^*(p)$ which was proved for two component stationary random media in [**24**] (see also [**25**]). The formulation there is in the continuum, but applies in the present context by replacing the continuum equations for the electric and current fields with their discrete analogs, i.e., Kirchoff's laws. We repeat here only the relevant features.

Let $s = 1/(1 - \epsilon)$. We shall consider s to be a complex variable. It can be shown that $\sigma^*(p, s)$ is analytic everywhere in the s-plane except for the interval $[0, 1]$. Furthermore, $\sigma^*(p, s)$ maps the upper half plane to the upper half plane, i.e., Im $\sigma^*(p,s) > 0$ when Im $s > 0$. As a consequence of these analytic properties, $\sigma^*(p,s)$ has the following integral representation,

$$1 - \sigma^*(p,s) = \int_0^1 \frac{d\mu(x)}{s-x} , \tag{4.1}$$

where μ is a positive Borel measure on $[0,1]$ which depends on p. Notice that this representation separates the dependence of $\sigma^*(p,s)$ on s from its dependence on p. (In fact, (4.1) applies even when $\epsilon = 0$.) The dependence of μ on p is most easily obtained through its moments, as follows. For $|s| > 1$, (4.1) can be expanded about a homogeneous medium ($s = \infty$ or $\epsilon = 1$), yielding

$$1 - \sigma^*(p,s) = \frac{\mu_0(p)}{s} + \frac{\mu_1(p)}{s^2} + \frac{\mu_2(p)}{s^3} + \dots , \tag{4.2}$$

$$\mu_n = \int_0^1 x^n d\mu(x) . \tag{4.3}$$

By equating (4.2) to a similar expansion of a resolvent representation for σ^*, one can obtain a formula for $\mu_n(p)$ in terms of the iterates of a self adjoint operator on L^2 (Ω = set of realizations of the bond conductivities) involving the Green's function of the discrete Laplacian. Because the bond conductivities are independent, these moments can be computed in principle (see, e.g., [**26**]), but they become very complicated. The first two are

$$\begin{aligned} \mu_0(p) &= 1 - p \\ \mu_1(p) &= \frac{p(1-p)}{d} . \end{aligned} \tag{4.4}$$

In general, $\mu_n(p)$ is an $(n+1)$-order polynomial in p.

We are now ready to state

LEMMA 4.1: *($d \geq 1$ bond problem) For every $\epsilon > 0$, there exists an open neighborhood V_ϵ in the complex p-plane such that $[0,1] \subset V_\epsilon$ and $\sigma^*(p)$ is analytic in V_ϵ.*

PROOF: Fix $s = 1/(1-\epsilon) > 1$. The idea is to produce a neighborhood containing $[0,1]$ in which (4.2) converges uniformly. Since for $p \in [0,1]$, $\mu_0(p) = 1 - p$ and $\mu_n(p) \geq \mu_{n+1}(p)$ for all n (via (4.3)),

$$\mu_n(p) \leq 1, \quad p \in [0,1] . \tag{4.5}$$

Now we must extend what we can of (4.5) into the complex plane. Consider $W = \{p \in \mathbb{C} | p \notin [0,1]\}$. Conformally map W onto the unit disk D in the z-plane, so that $p = \infty$ gets mapped to $z = 0$, and $[0,1]$ gets mapped to the unit circle $|z| = 1$. Let $m = n+1$. Since $\mu_n(p)$ is an mth order polynomial in p, $\mu_n(z)$ has at worst an mth order pole at $z = 0$. Thus $z^m \mu_n(z)$ is analytic in D. Since $|\mu_n(z)| \leq 1$ for $|z| = 1$, by the maximum modulus principle,

$$|\mu_n(z)| \leq \frac{1}{|z|^m} , \quad z \in D . \tag{4.6}$$

For any small $\delta' > 0$, there is a small $\delta > \delta' > 0$ such that in the annulus $A_{\delta'}$ defined by $1 \geq |z| > 1 - \delta'$

$$|\mu_n(z)| \leq (1+\delta)^m , \quad z \in A_{\delta'} . \tag{4.7}$$

For our given $s > 1$ (or $\epsilon > 0$), we can choose δ and δ' such that

$$|\mu_n(p)| \leq (1+\delta)^m < s^m , \quad p \in V_\epsilon , \tag{4.8}$$

where V_ϵ conformally maps to $A_{\delta'}$. Then (4.2) converges uniformly in V_ϵ, which proves the lemma.

The conformal mapping trick used to obtain (4.8) arose from a wonderful conversation with C. McMullen and C. Simpson, and the author gratefully acknowledges this.

Remark. Lemma 4.1 and its proof hold for a large class of continuum systems as well, namely infinitely interchangeable media, which have recently been introduced by O. Bruno [**27**]. This class is a generalization of Miller's cell materials [**28**], where all of space is divided up into cells, such as spheres of all sizes, and then the conductivity of each cell is a random variable (independent from the others) taking two (or more) values with probability p and $1-p$. While the integral representation (4.1) holds in great generality, along with (4.5), what is needed to make the proof go through is that the $\mu_n(p)$ are polynomials in p. The proof of this fact for infinitely interchangeable media is contained in [**29**] (along with rigorous upper and lower bounds on $\sigma^*(p)$ for the $d = 2$ bond problem with $\epsilon > 0$). We also note that Lemma 4.1 presumably does not hold for all composite media. For example, $\sigma^*(p)$ for a periodic array of spheres of volume fraction p embedded in a host material is believed to be analytic at $p = 0$ only in the variable $p^{\frac{1}{3}}$, so that $\sigma^*(p)$ has a branch cut there (see, e.g., [**30**]).

We now return to convexity. In the previous section it was found that convexity of $\sigma^*(p, \epsilon)$ appears to be lost only when ϵ becomes small enough. The following result provides some basis for this observation.

PROPOSITION 4.1: *($d \geq 1$ bond problem) For ϵ sufficiently close to 1, $\sigma^*(p)$ is convex for all $p \in [0,1]$.*

PROOF: From (4.2) and (4.4)

$$\sigma^*(p,s) = 1 - \left(\frac{1-p}{s} + \frac{p(1-p)/d}{s^2} + \ldots \right) . \tag{4.9}$$

By Lemma 4.1,

$$\frac{d^2\sigma^*}{dp^2} = \frac{2/d}{s^2} + O\left(\frac{1}{s^3}\right) , \tag{4.10}$$

which is positive for all $p \in [0,1]$ when s is sufficiently large.

In the previous section it was observed that convexity can only be lost near $p = 1$. At the other end, $p = 0$, convexity can never be lost, as shown in

PROPOSITION 4.2: *For any finite graph G with $\epsilon \geq 0$,*

$$\left.\frac{d^2\sigma}{dp^2}\right|_{p=0} \geq 0 . \tag{4.11}$$

PROOF: The proof is elementary and follows directly from (5.9).

Finally we come to the principal result of this section.

THEOREM 4.1: *($d = 2$ bond problem) For every $\epsilon > 0$, there exists an open neighborhood $\mathcal{U}_\epsilon \subset [0,1]$ containing $p_c = \frac{1}{2}$ such that $\sigma^*(p)$ is convex on $\mathcal{U}_\epsilon$.*

PROOF: The proof is based on Keller's Interchange Theorem

$$\sigma^*(\sigma_1, \sigma_2)\sigma^*(\sigma_2, \sigma_1) = \sigma_1\sigma_2 , \tag{4.12}$$

where $\sigma^*(\sigma_1, \sigma_2)$ is the bulk conductivity of a statistically isotropic, two-component stationary random medium in $d = 2$ with component conductivities σ_1 and σ_2, and $\sigma^*(\sigma_2, \sigma_1)$ is the bulk conductivity with σ_1 and σ_2 interchanged [**31-33**]. For the $d = 2$ bond lattice with $\sigma_1 = 1$ in proportion p and $\sigma_2 = \epsilon$ in proportion $1 - p$, (4.12) is also known as a duality relation [**16**], and is written as

$$\sigma^*(p)\sigma^*(1-p) = \epsilon , \tag{4.13}$$

so that at $p = \frac{1}{2}$

$$\sigma^*(\tfrac{1}{2}) = \sqrt{\epsilon} . \tag{4.14}$$

In order to prove the theorem, we compute the second derivative of $\sigma^*(p)$ at $p = \frac{1}{2}$,

$$\frac{d^2\sigma^*}{dp^2}|_{p=\frac{1}{2}} = \lim_{h\to 0} \frac{\sigma^*(\frac{1}{2}+h) - 2\sigma^*(\frac{1}{2}) + \sigma^*(\frac{1}{2}-h)}{h^2} \tag{4.15}$$

$$= \lim_{h\to 0} \frac{\sigma^*(\frac{1}{2}+h) - 2\sqrt{\epsilon} + \epsilon/\sigma^*(\frac{1}{2}+h)}{h^2} , \tag{4.16}$$

where (4.16) is obtained via (4.13) and (4.14). To compute the limit, we need some of the smoothness of $\sigma^*(p)$ provided by Lemma 4.1,

$$\sigma^*(\tfrac{1}{2}+h) = \sqrt{\epsilon} + \alpha h + O(h^2) , \tag{4.17}$$

where $\alpha > 0$ (when $0 < \epsilon < 1$). Inserting (4.17) into (4.16) and taking the limit, we obtain

$$\left.\frac{d^2\sigma^*}{dp^2}\right|_{p=\frac{1}{2}} = \frac{\alpha^2}{\sqrt{\epsilon}} > 0 . \tag{4.18}$$

Again using Lemma 4.1, $\frac{d^2\sigma^*}{dp^2}$ is a continuous function of p, so that it is positive in a neighborhood $\mathcal{U}_\epsilon$ of $p_c = \frac{1}{2}$, which proves the theorem.

Remark 1. It is interesting to note that $\frac{d\sigma^*}{dp^2}|_{p=\frac{1}{2}}$ depends only on $\frac{d\sigma^*}{dp}|_{p=\frac{1}{2}}$ and $\sigma^*(\frac{1}{2})$.

Remark 2. Our argument, while very simple, gives no control on the size of $\mathcal{U}_\epsilon$ as $\epsilon \to 0$, so that one cannot rigorously conclude from it alone that convexity is preserved in the limit. As stated before, however, it is known that the conductivity exponent for $d = 2$ satisfies $t \geq 1$, which indicates convexity near $p_c = \frac{1}{2}$ in the $\epsilon \to 0$ limit. In view of our result one can view the fact that $t \geq 1$ in $d = 2$ as arising from duality.

Remark 3. As noted above, Keller's Theorem holds in great generality, in particular, for infinitely interchangeable media in the continuum, as does Lemma 4.1. Thus Theorem 4.1 holds for infinitely interchangeable media as well. Information about the shape of $\sigma^*(p)$ in the insulator/conductor transition regime for small $\epsilon > 0$ is provided by the theorem only for those systems whose percolation threshold is known to occur at $p_c = \frac{1}{2}$. For percolating random systems in the continuum, apparently little is known about this. However, there is a periodic system which has an insulator/conductor transition occurring at $p_c = \frac{1}{2}$ and to which Theorem 4.1 presumably applies. This system is a variant of the one considered in a classical problem dating back to Maxwell [**34**] and Rayleigh [**35**], namely, a periodic array of perfectly conducting spheres occupying a volume fraction p of a host medium of unit conductivity. The variation that we consider here in $d = 2$ has been studied in [**36**] and [**37**], and is described as follows. At each integer pair $(i, j) \in \mathbb{Z}^2$ place a small square prism so that the center of the prism is at (i, j) and its vertices lie on the line segments joining (i, j) to its four nearest neighbors. Let the prisms have conductivity 1 and occupy a volume fraction p, and fill the rest of space with a medium of conductivity ϵ, with $0 < \epsilon << 1$. As the prisms grow in size, p increases. When $p = \frac{1}{2}$, the corners of the prisms touch, and it is around $p = \frac{1}{2}$ that the conducting transition occurs (when $\epsilon = 0$, $p_c = \frac{1}{2}$). Keller's Theorem applies to this problem, and presumably $\sigma^*(p)$ is analytic in p when $\epsilon > 0$, at least near $p = \frac{1}{2}$, in which case Theorem 4.1 applies as well, which says that $\sigma^*(p)$ is convex in the transition regime when $\epsilon > 0$. This finding is consistent with Figure 4 in [**36**]. The interesting question here is what happens as $\epsilon \to 0$. Is the convexity near $p = \frac{1}{2}$ a finite ϵ effect which vanishes as $\epsilon \to 0$, or does it persist as $\epsilon \to 0$. In the $\epsilon = 0$ limit, convexity is <u>not</u> a general feature of these types of systems near the transition. For a related problem in $d = 2$ involving circles (with $p_c \neq \frac{1}{2}$) rather than squares, the critical exponent for conductivity is $\frac{1}{2}$ [**38**], indicating concavity rather than convexity at threshold when $\epsilon = 0$.

5. Convexity when $\epsilon = 0$ and Conjectured Bounds on t. We begin this section by proposing a physical argument which we believe explains observed convexity of $\sigma^*(p)$ in bond and site lattice problems with $\epsilon = 0$ and $d \geq 2$ for p near p_c, $p > p_c$. Before giving the argument, we must introduce the notion of correlation length, which for $p > p_c$ is somewhat more delicate than for $p < p_c$. For the infinite bond lattice in $d \geq 2$ with a fraction p of occupied bonds, let

$$\tau^f(0, x) = \text{Prob}_p\{0 \text{ and } x \text{ belong to the same } \underline{\text{finite}} \text{ cluster of occupied bonds}\}\ . \tag{5.1}$$

Then the correlation length can be defined (see [**39**]) by

$$\frac{1}{\xi(p)} = \lim_{x \to \infty} -\frac{1}{|x|} \log \tau^f(0, x)\ , \tag{5.2}$$

where the limit is taken as x moves out to infinity in a fixed direction. This limit was proved to exist for $p > p_c$ in [**39**]. We shall assume that ξ diverges with exponent ν,

$$\xi(p) \sim (p - p_c)^{-\nu}, \quad p \to p_c^+\ . \tag{5.3}$$

For simplicity we formulate the argument for the $d = 2$ bond problem. Fix $p > p_c$ and consider an $L \times L$ sample of the square lattice with $L >> \xi(p)$. Let us view convexity as a decrease in $\frac{d\sigma^*}{dp}$ as p decreases, where for convenience we have dropped the L subscript in (2.2). To accomplish a decrease in $\frac{d\sigma^*}{dp}$, remove one occupied bond b_1 from the sample, and let $\delta_1\sigma$ be the expected drop in the conductivity as we average over all possible removals. It should be remarked that the occupied bonds can be divided into two types, the "backbone" and the "dangling" bonds. Backbone bonds have current flowing through them and dangling bonds do not – they are dead ends for the current. Consequently, a decrease in the conductivity can only be obtained by removing a backbone bond; the removal of a dangling bond does not contribute to $\delta_1\sigma$. Now, the removal of a backbone bond can create many new dangling bonds, for example if it breaks a "string" of connected bonds. Presumably, the probability of creating dangling bonds many correlation lengths away from b_1 is exponentially small. Consider then the removal of a second occupied bond b_2. If b_2 is far away (with respect to ξ) from b_1, then the contribution from such bonds to the expected drop $\delta_2\sigma$ in the conductivity will be essentially $\delta_1\sigma$. However, when b_2 is close to b_1, the corresponding average contribution to $\delta_2\sigma$ will be less than $\delta_1\sigma$, due to the increased density of dangling bonds around b_1, the removal of which contributes nothing to $\delta_2\sigma$. Thus $\delta_2\sigma < \delta_1\sigma$, which is equivalent to convexity. Far away from p_c the creation of dangling bonds will be a minor effect. For example, at $p = 1$ the removal of a single bond cannot create any dangling bonds. However as $p \to p_c^+$, ξ diverges, the backbone becomes more "stringy", and this effect will be more pronounced.

In order to make the intuition in the above argument more quantitative, we define the following quantities. Let G be any graph (we have in mind an $L \times L$ sample of the square lattice) with N bonds having conductivities 1 or $\epsilon \geq 0$, and let ω^k be as in (2.1). Now define

$$\delta\widetilde{\sigma}(\omega^k) = \sum_{i=1}^{N-k} \left[\widetilde{\sigma}(\omega_i^k(1)) - \widetilde{\sigma}(\omega_i^k(\epsilon))\right], \tag{5.4}$$

where given ω^k, i runs over the $N-k$ bonds which have conductivity 1, $\omega_i^k(1) = \omega^k$, and $\omega_i^k(\epsilon)$ is the same realization but with the ith bond conductivity changed to ϵ. Similarly, let

$$\delta^2\widetilde{\sigma}(\omega^k) = \sum_{\substack{i,j=1\\ i\neq j}}^{N-k} \left[\widetilde{\sigma}(\omega_{ij}^k(1,1)) + \widetilde{\sigma}(\omega_{ij}^k(\epsilon,\epsilon)) - \widetilde{\sigma}(\omega_{ij}^k(1,\epsilon)) - \widetilde{\sigma}(\omega_{ij}^k(\epsilon,1))\right], \tag{5.5}$$

where, given ω^k, i and j run over the $N-k$ bonds which have conductivity 1, $\omega_{ij}^k(1,1) = \omega^k$, $\omega_{ij}^k(\epsilon,\epsilon)$ is ω^k but with the ith and jth bond conductivities changed to ϵ, $\omega_{ij}^k(1,\epsilon)$ is ω^k but with the jth bond conductivity changed to ϵ, and similarly for $\omega_{ij}^k(\epsilon,1)$ with i instead of j. The expressions in (5.4) and (5.5) represent discrete first and second derivatives of the conductivity with respect to p. To make this connection more precise, we define

$$\beta_k = \sum_{\omega^k \in \Omega^k} \delta\widetilde{\sigma}(\omega^k) \tag{5.6}$$

$$\gamma_k = \sum_{\omega^k \in \Omega^k} \delta^2 \tilde{\sigma}(\omega^k) \,. \tag{5.7}$$

Then the exact relation is contained in

LEMMA 5.1: *Let G be any N-bond graph with bond conductivities 1 and $\epsilon \geq 0$. Then*

$$\frac{d\sigma}{dp} = \sum_{k=0}^{N-1} \beta_k p^{N-1-k} q^k \tag{5.8}$$

$$\frac{d^2\sigma}{dp^2} = \sum_{k=0}^{N-2} \gamma_k p^{N-2-k} q^k \,. \tag{5.9}$$

PROOF: Differentiate $\sigma(p)$ in (2.1) keeping in mind that since $q = 1-p$, $\frac{d}{dp} = -\frac{d}{dq}$. Then we have

$$\begin{aligned}
\frac{d\sigma}{dp} &= \sum_{k=0}^{N} \alpha_k \left[(N-k)p^{N-k-1}q^k - kp^{N-k}q^{k-1} \right] \\
&= \sum_{k=0}^{N-1} p^{N-k-1}q^k \left[(N-k)\alpha_k - (k+1)\alpha_{k+1} \right] ,
\end{aligned}$$

which can be written as (5.8). Taking one more derivative, we have

$$\begin{aligned}
\frac{d^2\sigma}{dp^2} &= \sum_{k=0}^{N} \alpha_k \Big[(N-k)(N-k-1)p^{N-k-2}q^k \\
&\qquad - 2(N-k)kp^{N-k-1}q^{k-1} + k(k-1)p^{N-k}q^{k-2} \Big] \\
&= \sum_{k=0}^{N-2} p^{N-2-k}q^k \Big[(N-k)(N-k-1)\alpha_k \\
&\qquad - 2(N-k-1)(k+1)\alpha_{k+1} + (k+2)(k+1)\alpha_{k+2} \Big] ,
\end{aligned}$$

which can be written as (5.9).

When the number of bonds in G goes to infinity, the appropriate limits of β_k and γ_k yield exact formulas for the first and second derivatives of $\sigma(p)$. When G is a hypercubic sample of the lattice with side L, then $N \sim dL^d$. In this case we consider $\sigma^*(p)$ in (2.3), assuming that the infinite volume limit exists. Then we have

LEMMA 5.2: *For the lattice in $d \geq 2$ with $\epsilon \geq 0$ and fixed p,*

$$\lim_{k,L\to\infty} L^{2-d}\beta_k/\binom{N}{k} = p\frac{d\sigma^*}{dp} \tag{5.10}$$

$$\lim_{k,L\to\infty} L^{2-d}\gamma_k/\binom{N}{k} = p^2\frac{d^2\sigma^*}{dp^2}\ , \tag{5.11}$$

where the simultaneous limits of k and $L \to \infty$ are taken so that

$$\lim_{k,L\to\infty}\frac{k}{N} = \lim_{k,L\to\infty}\frac{k}{dL^d} = q = 1-p\ .$$

PROOF: Equation (5.8) can be written as

$$p\frac{d\sigma}{dp} = \sum_{k=0}^{N-1}\beta_k p^{N-k}q^k\ . \tag{5.12}$$

Note that in (5.6) there are $\binom{N}{k}$ terms $\delta\tilde{\sigma}(\omega^k)$. When N is large, the weight of the binomial distribution is concentrated on values of k such that k/N is nearly $q = 1-p$. Appropriately scaling β_k with L^{2-d} yields the result (5.10), and similarly for (5.11). See [**40**] for more details about this type of argument.

It is important at this juncture to point out the implications of Lemmas 5.1 and 5.2 for our analysis of $\frac{d^2\sigma^*}{dp^2}$. For simplicity, let us consider $d = 2$ (with $\epsilon = 0$) so that $L^{2-d} = 1$ in (5.11), which can be written as (after dropping the "*" notation)

$$p^2\frac{d^2\sigma}{dp^2} = <\delta^2\tilde{\sigma}(\omega^p)>_p\ , \tag{5.13}$$

where ω^p is a configuration of the bond lattice with a fraction p of the bonds occupied (which can be viewed as a random graph), and $<\cdot>_p$ denotes averaging over such configurations. Since $\delta^2\tilde{\sigma}(\omega^p)$ involves all pairs of bonds in ω^p, presumably an ergodic theory argument shows that it suffices in (5.13) to consider a single "typical" configuration $B(p)$ of the bond lattice with a fraction p of occupied bonds (i.e., $B(p)$ belongs to a set of full Bernoulli measure in the standard percolation problem). Then (5.13) can be written as

$$p^2\frac{d^2\sigma}{dp^2} = \delta^2\sigma(B(p))\ , \tag{5.14}$$

where we have now dropped the "~" notation. Also note that, given any graph G with bond conductivities 1 and 0 and conductivity function $\sigma(p)$,

$$\left.\frac{d^2\sigma}{dp^2}\right|_{p=1} = \delta^2\sigma(G)\ . \tag{5.15}$$

Then we see via (5.14) that $p^2 \frac{d^2\sigma}{dp^2}(p)$ for the lattice can be computed by finding $\frac{d^2\widehat{\sigma}}{dp^2}|_{p=1}$, where $\widehat{\sigma}$ is the conductivity function for $G = B(p)$. Similar considerations hold for $\frac{d\sigma}{dp}$, with the analog of (5.15) being

$$\left.\frac{d\sigma}{dp}\right|_{p=1} = \delta\sigma(G)\ . \tag{5.16}$$

We now return to the physical argument given above. The principal objection that one can raise to this argument is that the first removal b_1 can set up a situation where the second removal results in a larger drop in the conductivity than did the first. For example, consider a configuration of connected bonds in the shape of a Y, where the current flows in through the single leg and out through the two arms. For simplicity let each "limb" of the Y be composed of 1 bond of conductivity 1, with b_1 the leg and b_2 and b_3 the arms. Consider now $\delta^2\sigma(Y) = 2(\delta_{12}+\delta_{23}+\delta_{13})$, where the δ_{ij} are the summands in (5.5). Elementary calculation shows that $\delta_{12} = \delta_{13} = +\frac{1}{6}$ and $\delta_{23} = \frac{-1}{3}$, so that $\delta^2\sigma(Y) = 0$. The important point to note is that $(2,3)$ is a pair of bonds in parallel, while $(1,2)$ and $(1,3)$ are pairs in series. The pair $(2,3)$ is an example of the objection raised. When b_2 is removed, the current can still flow through b_3, and the drop in conductivity is minimal. When b_3 is subsequently removed, the effect is to cut off all current flow, with the result that the net contribution of the pair to $\delta^2\sigma(Y)$ is negative. However, when b_1 is removed first, the flow is stopped immediately, so that b_2 and b_3 are dangling bonds whose subsequent removal does not affect the conductivity, which is the principal reason why δ_{12} and δ_{13} are positive.

For an arbitrary graph G, there is no particular reason why the positive δ_{ij}'s should outweigh the negative ones. However, for graphs that are sufficiently stringy, positive contributions from series pairs in a given string should tip the balance to a net positive $\delta^2\sigma(G) > 0$. For example, if we replace each bond in Y above with 2 bonds in series, and call the new graph Y_2, then $\delta^2\sigma(Y_2) > 0$. The reason is that the three series pairs in the leg and two arms give new, positive contributions to $\delta^2\sigma(Y_2)$ which were not present in $\delta^2\sigma(Y)$. Such considerations led us to the following

THEOREM 5.1: *Let G be any finite connected graph with bond conductivities 1 or 0, and let S_nG be a new graph formed by replacing each bond of G with n bonds in series. Then*

$$\delta^2\sigma(S_nG) = n\delta^2\sigma(G) + (n-1)\delta\sigma(G) \tag{5.17}$$

$$\delta\sigma(G) \geq 0\ . \tag{5.18}$$

Before we prove the theorem, its principal consequence is given in

COROLLARY 5.1: *If G satisfies*

$$|\delta^2\sigma(G)| < \delta\sigma(G)\ , \tag{5.19}$$

then there exists a positive constant C such that

$$\delta^2\sigma(S_nG) \sim Cn\,, \quad C > 0\;,\; n \to \infty\;. \tag{5.20}$$

PROOF 1 OF THEOREM 5.1: In order to calculate $\delta^2\sigma(S_nG)$, we must consider all pairs of bonds (i,j), $i \neq j$, in the new graph S_nG. Let the pairs in the original graph be labeled by (i',j'), $i' \neq j'$. Now, the pairs in S_nG are of two types, those which arise from $i' \neq j'$ and those which arise from one original bond i'. Denoting the summands in (5.5) again as δ_{ij}, for those (i,j) pairs of the first type, we have

$$\delta_{ij} = \frac{1}{n}\delta_{i'j'}\;, \tag{5.21}$$

where the factor of $\frac{1}{n}$ appears because

$$\sigma(S_nG) = \frac{1}{n}\sigma(G)\;, \tag{5.22}$$

since the conductivity of n bonds in series is $\frac{1}{n}$. For pairs of the second type,

$$\delta_{ij} = \frac{1}{n}\left(\sigma_{i'}(1) - \sigma_{i'}(0)\right)\;, \tag{5.23}$$

where $\sigma_{i'}(1) = \sigma(G)$ and $\sigma_{i'}(0)$ is σ of G without bond i'. Note that δ_{ij} in (5.23) is always non-negative. For each (i',j') pair in G, $i' \neq j'$, there are n^2 pairs of the first type in S_nG. For each i' in G, there are $n(n-1)$ pairs of the second type in S_nG. Thus

$$\delta^2\sigma(S_nG) = \frac{1}{n}\left(n^2\delta^2\sigma(G) + n(n-1)\delta\sigma(G)\right)\;, \tag{5.24}$$

which yields (5.17).

Remark. S_nG is composed of a graph G whose elements are strings of n bonds in series. Removal of any one of these bonds converts all the other bonds in that string into dangling bonds. Their subsequent removal has no effect on the conductivity, and this effect is the source of the positive term (5.23), which becomes the second term in (5.17).

PROOF 2 OF THEOREM 5.1: Let the conductivity function of S_nG be denoted by $\widehat{\sigma}(p)$, and that of G be denoted by $\sigma(p)$. Then

$$\widehat{\sigma}(p) = \frac{1}{n}\sigma(p^n)\,. \tag{5.25}$$

Differentiating both sides of (5.25) twice with respect to p and setting $p = 1$ yields

$$\left.\frac{d^2\widehat{\sigma}}{dp^2}\right|_{p=1} = n\left.\frac{d^2\sigma}{dp^2}\right|_{p=1} + (n-1)\left.\frac{d\sigma}{dp}\right|_{p=1}\;, \tag{5.26}$$

which is equivalent to (5.17).

Proof 2 was a joint observation with S. Goldstein, which was made subsequent to the original Proof 1. We chose to include Proof 1 as well because it shows explicitly the convexity improving effect of creating dangling bonds.

We are interested in applying Theorem 5.1 and its corollary to the lattice. Henceforth let G_L be a square (or cubic or hypercubic) sample of side L of the square (or cubic or hypercubic) lattice in $d \geq 2$ with $p = 1$. (Think of G_L as standing for "grid".) Let G be the infinite volume limit of G_L. Furthermore, let

$$\delta\sigma^*(G) = \lim_{L\to\infty} L^{2-d}\delta\sigma_L(G_L) \tag{5.27}$$

$$\delta^2\sigma^*(G) = \lim_{L\to\infty} L^{2-d}\delta^2\sigma_L(G_L)\ , \tag{5.28}$$

where on the right hand sides of (5.27) and (5.28) we have simply added an L subscript to the notation used in (5.15) and (5.16). In order for Corollary 5.1 to apply to G_L and G, condition (5.19) (appropriately scaled in L) must be satisfied. As stated before, for $d = 2$,

$$\delta\sigma^*(G) = \frac{d\sigma^*}{dp}\Big|_{p=1} = 2 \tag{5.29}$$

$$\delta^2\sigma^*(G) = \frac{d^2\sigma^*}{dp^2}\Big|_{p=1} \approx -0.21\ , \tag{5.30}$$

so that condition (5.19) is presumably satisfied for both G and G_L with L large, although we have not rigorously proven such a statement. In higher dimensions [**7**]

$$\delta\sigma^*(G) = \frac{d\sigma^*}{dp}\Big|_{p=1} = \frac{d}{d-1}\ , \tag{5.31}$$

while numerical simulation of $\sigma^*(p)$ in $d = 3$ [**6**] and analytical solution of $\sigma^*(p)$ for the Bethe lattice, supposedly representing large d [**41, 42**], are practically linear near $p = 1$, so that $\delta^2\sigma^*(G)$ is also small, as in $d = 2$. We thus state an unproven

CONJECTURE 5.1: *Condition (5.19) is satisfied by the bulk conductivity σ^* for the square (or cubic or hypercubic) lattice in $d \geq 2$.*

Let us now describe the picture we have in mind. We begin with G_L, and apply S_n to it, for some large n. The result, S_nG_L, can be thought of as a super lattice or grid with side of length nL, composed of "strings" or "macrolinks" connecting the old vertices or "nodes" of G_L. This super lattice is closely connected to the so-called "node-link" model of the backbone of the infinite cluster for p near p_c, $p > p_c$, proposed independently by Skal and Shlovskii [**43**] and de Gennes [**44**]. In this model, the length n of the strings connecting the nodes is greater than the distance between the nodes, reflecting the observation in actual clusters that "strings" are not straight but tend to wander around. The distance between the nodes is assumed to be on the order of a correlation length ξ. The geometrical parameters n and ξ in the node-link model are related to p via the scaling relations as $p \to p_c^+$,

$$\xi(p) \sim (p - p_c)^{-\nu} \tag{5.32}$$

$$n(p) \sim (p - p_c)^{-\zeta} \ . \tag{5.33}$$

The Chayes [12] have rigorously shown that $\zeta \geq \nu$, and also that $\zeta \geq \min\{1, \frac{\nu}{\nu'}\}$, for an appropriately defined ζ, where ν' is the correlation length exponent as p approaches p_c from below. It is believed, however, that $\nu' = \nu$, and we shall assume that $\zeta \geq 1$. Under these assumptions, an easy calculation leads to the following expression for the conductivity critical exponent,

$$t = (d - 2)\nu + \zeta \ . \tag{5.34}$$

We can make the correspondence between our super lattice $S_n G_L$ and the node-link model exact by allowing our strings to wander as well, and stipulating that the length of a side of $S_n G_L$ is ξL, rather than nL. This variation does not alter the conductivity or its derivatives, but only the way the graph is situated in space. It should be remarked that in the node-link model we have generated via $S_n G_L$, four strings meet at each node in $d = 2$. Apparently, though, it is much more common in actual percolation clusters to observe three fold meetings (D. Fisher, private communication). This can be taken into account in our model by letting G_L be a sample of the hexagonal lattice instead of the square lattice. Presumably (5.19) still holds for the hexagonal lattice. Similar considerations apply in higher dimensions as well.

Apparently it is now generally accepted [14, 12] that the node-link model is an oversimplification of the backbone structure, particularly in low dimension, especially $d = 2$. Stanley [45] has suggested that a more accurate representation of the backbone is provided by a "node-link-blob" model. In this model, nodes separated by a distance ξ are connected by strings or links (of singly connected bonds) and blobs (of multiply connected bonds). One can visualize the connection between nodes as a segment of a necklace of beads on a string where there is some distance between the beads. Also, a node may actually be a blob. These blobs have a self-similar structure, i.e., they have a node-link-blob structure themselves.

We now wish to explore the consequences of Theorem 5.1 and its corollary, under the assumption of Conjecture 5.1 The case of $d = 2$ is considered first. Theorem 5.1 yields

$$\delta^2\sigma(S_n G_L) = n\delta^2\sigma(G_L) + (n-1)\delta\sigma(G_L) \ . \tag{5.35}$$

Since $L^{2-d} = 1$ for $d = 2$, we can directly take the infinite volume limit of (5.35) (assuming it exists) to obtain

$$\delta^2\sigma^*(S_n G) = n\delta^2\sigma^*(G) + (n-1)\delta\sigma^*(G) \ . \tag{5.36}$$

Under Conjecture 5.1,

$$\delta^2\sigma^*(S_nG) \sim C_2 n\,, \quad n \to \infty\,, C_2 > 0\;. \tag{5.37}$$

The upshot of (5.37) is that, if one assumes that the backbone of the infinite cluster behaves like the node-link model, then not only is $\frac{d^2\sigma^*}{dp^2}$ positive as $p \to p_c^+$, indicating convexity, but it diverges to $+\infty$. Presumably a similar result can be obtained for a good approximation of a node-link-blob model by taking an appropriate G with a sufficient number of levels of self similarity in each blob (and such that (5.19) holds). Such considerations lead us to

CONJECTURE 5.2: *($d = 2$ bond problem)*

$$\frac{d^2\sigma^*}{dp^2} \to +\infty \quad \text{as } p \to p_c^+\;. \tag{5.38}$$

Consequently, $\sigma^*(p)$ is convex in $(p_c, p_c + a)$, for some small a, and

$$1 \le t < 2\;. \tag{5.39}$$

The first inequality in (5.39), as already mentioned, has been rigorously established, but is explained here via convexity of σ^* near $p_c = \frac{1}{2}$. The second inequality $t < 2$ comes from (5.38) and

$$\frac{d^2\sigma^*}{dp^2} \sim (p - p_c)^{t-2}\;. \tag{5.40}$$

A tighter upper bound on t is provided by the node-link relation for t in (5.34) for $d = 2$,

$$t \le \zeta\,, \tag{5.41}$$

where a reasonable numerical estimate for ζ in $d = 2$ is about 1.35 [**46**]. A rigorous argument yielding (5.41) was shown to the author by H. Kesten. What this bound amounts to is that the conductivity of the node link backbone is smaller than the conductivity of the actual backbone, i.e.,

$$\sigma^*(S_nG) \le \sigma^*\big(B(p)\big)\;, \tag{5.42}$$

where $B(p)$ is a typical backbone configuration at volume fraction p, and n is given by (5.33). Inequality (5.42) is physically reasonable if we imagine $B(p)$ to be constructed from S_nG by adding bonds, which increases the conductivity (see also [**12**]).

We now apply Theorem 5.1 to higher dimensions, with $d = 3$ first. Dividing (5.35) by the length ξL of a side of S_nG_L yields

$$\frac{\delta^2\sigma(S_nG_L)}{\xi L} = \frac{n\delta^2\sigma(G_L)}{\xi L} + \frac{(n-1)\delta\sigma(G_L)}{\xi L}\;. \tag{5.43}$$

Taking the limit as $L \to \infty$ gives

$$\delta^2\sigma^*(S_nG) = \frac{n}{\xi}\delta^2\sigma^*(G) + \frac{n-1}{\xi}\delta\sigma^*(G) \ . \tag{5.44}$$

Under Conjecture 5.1,

$$\delta^2\sigma^*(S_nG) \sim C_3\frac{n}{\xi}, \quad n, \xi \to \infty, C_3 > 0 \ . \tag{5.45}$$

Now, as we will discuss below, we have rigorous evidence that

$$\delta^2\sigma^*(S_nG) \geq \delta^2\sigma^*\big(B(p)\big) \ , \tag{5.46}$$

which, using (5.45), (5.32) and (5.33), becomes

$$\frac{d^2\sigma^*}{dp^2} \sim (p - p_c)^{t-2} \leq (p - p_c)^{\nu-\zeta} \ , \tag{5.47}$$

giving a lower bound on t. Repeating the same procedure, but dividing by the appropriate power of ξL in (5.43), and considering the analog of (5.42) in higher dimensions, leads us to

CONJECTURE 5.3: *($3 \leq d \leq 6$ bond problem)*

$$(d-2)\nu + (2-\zeta) \leq t \leq (d-2)\nu + \zeta, \quad \zeta \geq 1 \ . \tag{5.48}$$

These bounds become tighter with increasing dimension and converge in $d = 6$ where it is believed that $\nu = \frac{1}{2}$ and $\zeta = 1$, with $t = 3$. Using numerical values for ζ and ν in dimensions 2 through 5 [**46, 47, 8**], we have plotted in Figure 1 the proposed bounds (5.48), as well as the bound $1 \leq t \leq \zeta$ in $d = 2$. In our choices of values for ζ, we have chosen the largest reasonable ones, which make the bounds the widest. The best current numerical estimate for t in $d = 2$ appears to be $t = 1.303^{+0.004}_{-0.014}$ [**48**], while in $d = 3$ the situation is not as well established. In [**49**], a "relatively well established" value of ≈ 1.9 is quoted, while very recent numerical estimates of C. Lobb indicate that t in $d = 3$ is "very close to 2" [C. Lobb, private communication]. These values fall within the proposed bounds. The only recent numerical estimates of t for $d = 4$ and 5 known to the author are those in [**46**] and [**47**], where t is computed via (5.34), which is just the upper bound in (5.48).

We now explain (5.46), which says that $\delta^2\sigma^*$ for the node-link model is greater than $\delta^2\sigma^*$ for an actual backbone configuration. The intuitive reason is fairly simple. The node-link model is composed purely of strings, and pairs of bonds from within a given string give purely positive contributions to $\delta^2\sigma^*$. However, the actual backbone is composed of strings <u>and</u> regions of multiply connected bonds. Imagine a blob composed of a piece of the lattice G at $p = 1$ (a "full" blob). Pairs of bonds from within the blob will give a small negative contribution to $\delta^2\sigma^*$. Even if the blob has some string structure within it, the net contribution to $\delta^2\sigma^*$ will still be less than if there were in its place a pure string with length of the same order as the size of the blob. Such statements can be proved under certain conditions, which provides a rigorous basis for (5.46) and forms the content of the informally stated

THEOREM 5.2: *Let $S(n)$ be a graph composed of n bonds in series, i.e., $S(n)$ is a string of length n. Let $T(n)$ be a "necklace" composed from $S(n)$, i.e., replace some sections of $S(n)$ with blobs, which can be pieces of the lattice G at $p = 1$, or pieces of $S_m G$, with m sufficiently small compared to the length of the section of $S(n)$ the blob replaced. Then*

$$\delta^2\sigma(S(n)) \geq \delta^2\sigma(T(n)) \ . \tag{5.49}$$

The proof of this theorem follows along the lines of Proof 1 of Theorem 5.1, but will be omitted here.

We close by remarking that Straley [**13**] has proposed a (non-rigorous) lower bound on t that is better than the one in (5.48), namely, $t \geq (d-2)\nu + 1$. However, our analysis leading to the lower bound in (5.48) has some interesting consequences for the behavior of $\frac{d^2\sigma^*}{dp^2}$ as $p \to p_c^+$, particularly in $d \geq 3$, which is discussed in [**50**].

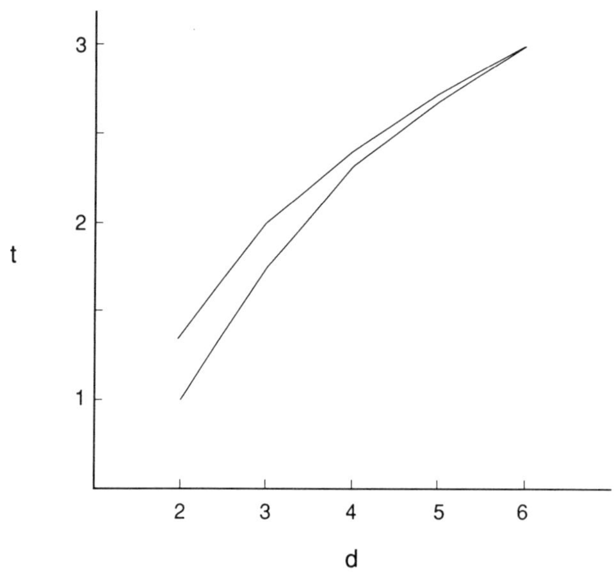

Figure 1. Proposed upper and lower bounds on the conductivity exponent t in terms of the percolation exponents ν and ζ. Numerical values for ν and ζ are used to evaluate the bounds in dimensions $d = 2, 3, 4, 5,$ and 6. Straight lines have been drawn between these points.

Acknowledgements. It is a pleasure to thank a number of people for enjoyable and enlightening discussions during the course of the work: O. Bruno, J. and L. Chayes, D. Fisher, S. Goldstein, H. Kesten, C. McMullen, G. Papanicolaou, C. Simpson, and S. Varadhan. Special thanks is owed to P. Doyle for many long and very useful discussions which clarified certain matters in this work. The author is also very grateful to N. Wilcox for persevering through the difficulties of typing a camera ready copy.

REFERENCES

1. J.M. Ziman, J. Phys. C, 1 (1968), p. 1532.
2. T.P. Eggarter and M.H. Cohen, Phys. Rev. Lett., 25 (1970), p. 807.
3. T.P. Eggarter and M.H. Cohen, Phys. Rev. Lett., 27 (1971), p. 129.
4. S.R. Broadbent and J.M. Hammersly, Proc. Cambridge Philos. Soc., 53 (1957), p. 629.
5. B.J. Last and D.J. Thouless, Phys. Rev. Lett., 27 (1971), p. 1719.
6. S. Kirkpatrick, Phys. Rev. Lett., 27 (1971), p. 1722.
7. S. Kirkpatrick, Rev. Mod. Phys., 45 (1973), p. 574.
8. D. Stauffer, Phys. Rep., 54 (1979), p. 1.
9. G. Grimmett and H. Kesten, Z. Wahr., 66 (1984), p. 335.
10. H. Kesten, Percolation Theory for Mathematicians, Birkhaüser, Boston, 1982.
11. J.T. Chayes and L. Chayes, Comm. Math. Phys., 105 (1986), p, 133.
12. J.T. Chayes and L. Chayes, Comm. Math. Phys., 113 (1987), p. 27.
13. J.P. Straley, Phys. Rev. C, 15 (1982), p. 2333.
14. A. Coniglio, J. Phys. A, 15 (1982), p. 3829.
15. L. Russo, Z. Wahr., 56 (1981), p. 229.
16. J.P. Straley, Phys. Rev. B, 15 (1977), p. 5733.
17. B.P. Watson and P.L. Leath, Phys. Rev. B, 9 (1974), p. 4893.
18. A.B. Harris and S. Kirkpatrick, Phys. Rev. B, 16 (1977), p. 542.
19. M.H. Ernst, P.F.J. van Velthoven, and Th. M. Nieuwenhuizen, J. Phys. A, 20 (1987), p. 949.
20. Th. M. Nieuwenhuizen, P.F.J. van Velthoven, and M.H. Ernst, Phys. Rev. Lett. 57 (1986), p. 2477.
21. R. Künnemann, Effective conductivity on a lattice as a limit of box conductivities, in Ph.D. Thesis, U. Of Heidelberg, 1983.
22. S.J. Mason and H.J. Zimmermann, Electronic Circuits, Signals, and Systems, John Wiley and Sons, New York, 1960.
23. C. Thomassen, Resistances and currents in infinite electrical networks, to appear in J. Comb. Th. B.
24. K. Golden and G. Papanicolaou, Comm. Math. Phys., 90 (1983), p. 473.
25. D.J. Bergman, Phys. Rep., C43 (1978), p.377.
26. D.J. Bergman and Y. Kantor, J. Phys. C, 14 (1981), p. 3365.
27. O. Bruno, The Effective Conductivity of an Infinitely Interchangeable Mixture, Ph.D. Thesis, New York University, 1989.
28. M.N. Miller, J. Math. Phys., 10 (1969), p. 1988.
29. O. Bruno and K. Golden, Interchangeability and bounds for the effective conductivity of the square lattice, in preparation.
30. A.S. Sangani and A. Acrivos, Proc. R. Soc. Lond. A, 386 (1983), p. 263.
31. J.B. Keller, J. Math. Phys., 5 (1964), p. 548.
32. A.M. Dykhne, Sov. Phys. JETP, 32 (1971), p. 63.

33. W. Kohler and G. Papanicolaou, Bounds for the effective conductivity of random media, in Lecture Notes in Physics Vol. 154, Springer Verlag, New York, 1982, p. 111.
34. J. C. Maxwell, A Treatise on Electricity and Magnetism, Clarendon Press, Oxford, 1892.
35. Lord Rayleigh, Phil. Mag., 34 (1892), p. 481.
36. G. W. Milton, R. C. McPhedran, and D. R. McKenzie, Appl. Phys., 25 (1981), p. 23.
37. R. Fogelholm and G. Grimvall, J. Phys. C, 16 (1983), p. 1077.
38. J. B. Keller, J. Appl. Phys., 34 (1963), p. 991.
39. J. T. Chayes, L. Chayes, G. R. Grimmett, H. Kesten, and R. H. Schonmann, The correlation length for the high density phase of Bernoulli percolation, to appear in Ann. Prob.
40. J. Lamperti, Probability, W. A. Benjamin, Inc., Reading, 1966.
41. R. B. Stinchcombe, J. Phys. C, 6 (1973), p. L1.
42. J. P. Straley, J. Phys. C, 10 (1977), p. 3009.
43. A. S. Skal and B. I. Shlovskii, Sov. Phys.- Semicond., 8 (1975), p. 1029.
44. P. G. de Gennes, J. Physique Lett., 37 (1976), p. L1.
45. H. E. Stanley, J. Phys. A, 10 (1977), p. L211.
46. J. Adler, J. Phys. A, 18 (1985), p. 307.
47. R. Fisch and A. B. Harris, Phys. Rev. B, 18 (1978), p. 416.
48. D. J. Frank and C. J. Lobb, Phys. Rev. B, 37 (1988), p. 302.
49. S. Feng, B. I. Halperin, and P. N. Sen, Phys. Rev. B, 35 (1987), p. 197.
50. K. Golden, Convexity and exponent inequalities for conduction near percolation, in preparation.

CHAPTER 10

Simulations of Composites: From Ohmic Response to the Breakdown of Superconductivity*

C. J. Lobb†

Abstract. Two fast algorithms for reducing random lattices to a single equivalent bond are described. One such algorithm can be used on two-dimensional lattices if the elements which make up the bonds of the lattice have a Y-∇ transformation. The second algorithm is not restricted to two dimensions, but requires that nodes can be removed from the lattice at the expense of increasing the interconnectedness of the remaining nodes. The algorithms are used to calculate the bulk conductance of large square and cubic percolative lattices of conductors. The two-dimensional algorithm can also be applied to some nonlinear problems, and is used here to study the current-induced breakdown of superconductivity in percolative lattices.

1. Introduction. Only a limited amount is known analytically about the bulk properties of composites, so that numerical methods are very important in this field. This paper will describe two numerical methods for finding the bulk conductance of random resistor networks [1-3]. Both of these methods are generalizations of the elementary method for finding

*Supported in part by NSF grants DMR84-04489, DMR86-14003 and INT87-14947.

†Division of Applied Sciences and Department of Physics, Harvard University, Cambridge, MA 02138.

the conductance of simple circuits: the circuit is repeatedly transformed, using series, parallel, and other rules, until an equivalent single conductance is all that remains.

In section 2, the basic transformations will be discussed. In section 3, they will be applied to reducing a square lattice to a single conductance. In section 4, results of applying the two-dimensional algorithm to percolative lattices are reviewed, and critical exponents for this problem are discussed. In section 5, two-dimensional percolative lattices of superconductors are discussed; the method is used to calculate the superconducting properties of such lattices near their percolation threshold. Finally, in section 6, a three-dimensional transformation algorithm is discussed, and preliminary results for three-dimensional percolation are presented.

2. Conductance transformations. A variety of transformations exist which allow one to modify the internal structure of circuits without changing their overall properties. Consider, for example, two conductances in series, G_1 and G_2. As is well known, and as illustrated in Figure 1(A), two conductors in series can be

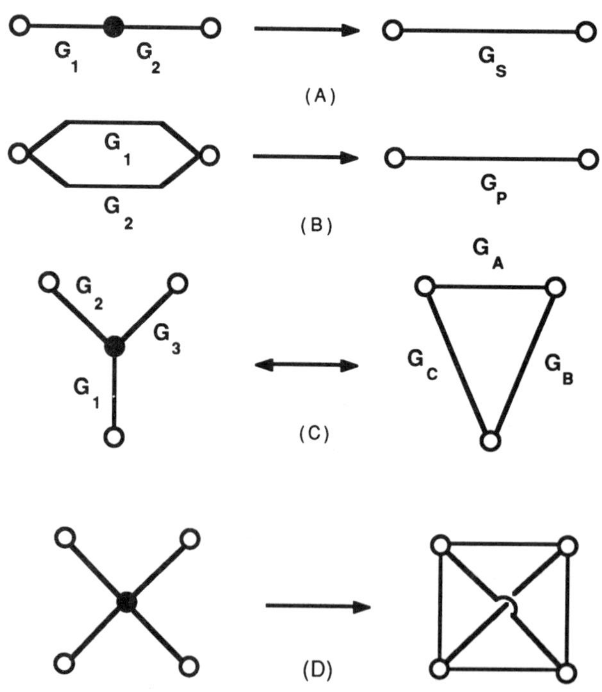

Figure 1. Transformations of conductances. (A)-(D) are series, parallel, Y-∇, and a general node-removal transformation.

combined to give

$$G_s = G_1 G_2 / (G_1 + G_2) \tag{1}$$

while two conductors in parallel, Figure 1(B), can be combined to give

$$G_p = G_1 + G_2 \quad . \tag{2}$$

In both cases, the resulting single conductor is completely equivalent to the two original conductors as far as any external circuit is concerned.

Also well known is the Y-∇ transformation [4], shown in Figure 1(C). This transformation is defined in both directions, the conductances given as follows:

$$Y \to \nabla: \qquad G_A = \frac{G_2 G_3}{G_1 + G_2 + G_3} \tag{3}$$

plus cyclic permutations, and

$$\nabla \to Y: \qquad G_1 = G_B G_C \left(\frac{1}{G_A} + \frac{1}{G_B} + \frac{1}{G_C} \right) \tag{4}$$

plus cyclic permutations. Note again that these transformations do not change the electrical properties measured through any external circuit.

A generalization of the $Y \to \nabla$ transformation is illustrated in Figure 1(D). It is posssible to remove any node from a conductor network, such as the dark node shown in the left-hand graph, at the expense of introducing conductances which connect every pair of nodes which were formerly connected to the removed node. The formula for the conductance of the new bonds is a simple generalization of (3), where the denominator becomes the sum of all of the conductances in the left-hand diagram, and the numerator is the product of the two conductances which formerly connected two nodes to the removed node. It is possible to have any number of conductances connected to the dark node and still do this transformation.

3. Square lattices. An additional transformation needed to reduce a square lattice to a single equivalent bond is shown in Figure 2. By combining $Y \to \nabla$ and $\nabla \to Y$ transformations, one can "propogate" a diagonal bond through a lattice. From Figure 2(A) to Figure 2(B), the upper ∇ is converted to a Y. The resulting Y consisting of the lower three bonds, re-drawn for clarity in Figure 2(C), is then transformed to a ∇ in Figure 2(D). The net effect of this propogation transformation is to move the diagonal bond, while changing the values of all five conductances.

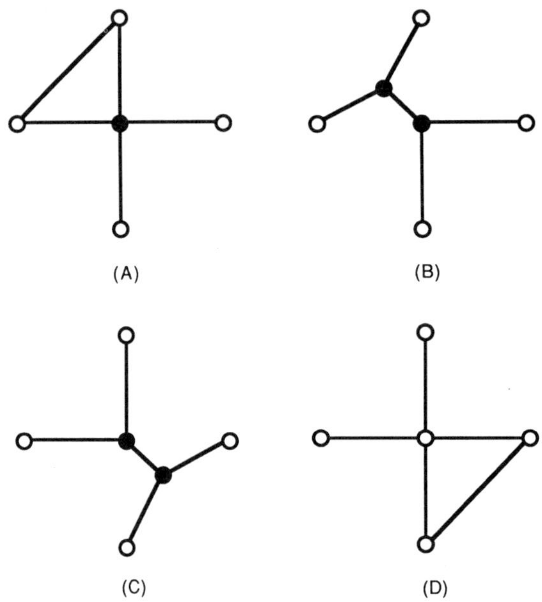

Figure 2. The bond propogation transformation. A $\nabla - Y$ transformation [(A) to (B)] followed by a Y-∇ transformation [(C) to (D)] changes the value of all five conductances shown while moving the diagonal through the lattice.

The propogation transformation is used in Figure 3 to reduce a small square lattice. First, the upper left-hand corner bonds are combined in series, going from Figure 3(A) to Figure 3(B). The resulting diagonal bond is then propogated through the lattice, as shown. This leaves two new corners, on the left, which may each be combined in series to go from Figure 3(C) to 3(D). The two resulting diagonal bonds are then separately propogated across the lattice. This process is repeated until a series chain of conductors remains, which are combined by the series rule into a single conductance. Details of this method, and a discussion of its efficiency, are published elsewhere [1]. As rough benchmarks, a lattice containing ~10^6 conductors takes about 30 seconds to solve on an IBM 3081, while smaller lattices containing ~10^4 conductors can be solved on a Macintosh in about 10 minutes.

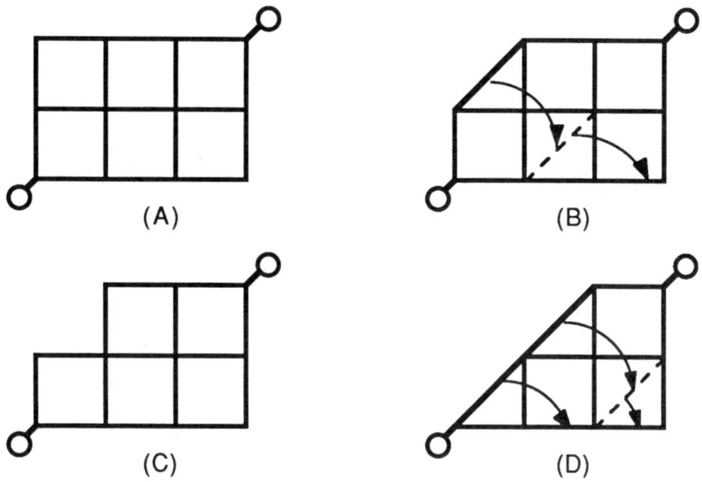

Figure 3. Solution of a small square lattice using series and bond propogation transformations.

4. Percolative conduction in two dimensions. As a specific example, consider a lattice consisting of bonds which have conductance 1 with probability p, and 0 with probability 1 - p. If the lattice has equipotentials attached to its left and right sides, and is infinite, the conductance G measured between the equipotentials will be zero up to the percolation threshold p_c. Just above threshold, the bulk conductance has a power-law dependence,

$$G \sim (p - p_c)^t \tag{5}$$

where t is the critical exponent for conductance.

The formula (5) can be used to estimate t by solving large lattices at various p near p_c, and then curve fitting. The disadvantage of this approach is that for samples smaller than the correlation length ξ, which diverges near p_c as

$$\xi \sim (p - p_c)^{-\nu} \tag{6}$$

(where ν = 4/3 in two dimensions [5]), (5) is no longer valid. Close to p_c, there are large sample-to-sample variations in the conductance of finite samples.

To avoid these problems, a finite-size scaling analysis can be done [6,7]. If b is the size of the sample measured in units of the lattice spacing, then the average bulk conductance $<G>$ at $p = p_c$ is given by

$$<G> = b^{-(2-d+t/\nu)} [c_1 + c_2 f_2(b) + ...] \tag{7}$$

where $f_2(b)$ approaches 0 as b approaches infinity, d is the spatial dimension of the lattice, and c_1 and c_2 are constants.

The formula (7) can be used to calculate t numerically [1,2]. $<G>$ has been calculated from $b = 2$ to $b = 1000$ by generating different realizations of the various size lattices and keeping track of arithmetic, geometric, and harmonic means. Figure 4 is a plot of this data, where the axes have been chosen so that the y-intercept of the data is t/ν, as can be seen by examining (7). The data are derived from every distinct realization for $b = 2$ and $b = 3$, 2×10^6 realizations for $b = 4$, continuing to 400 realizations for $b = 1000$. The curves are independent fits to the different means using (7) with $f_2(b) = b^{-\Delta/\nu}$. The value of t thus determined is [1]

$$t = 1.303 \quad (+0.004/-0.014) \quad . \tag{8}$$

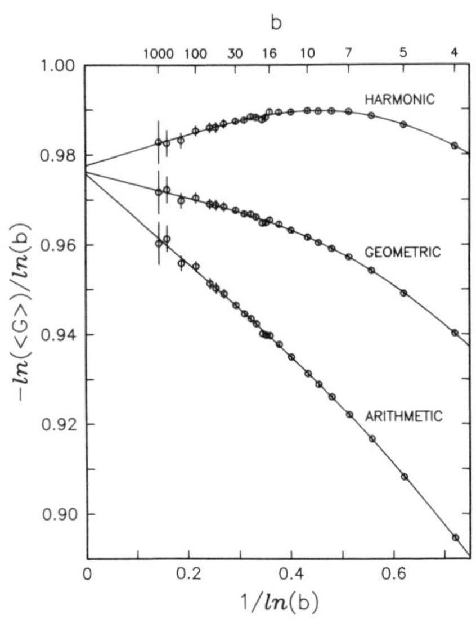

Figure 4. Plot of data from square lattices. The mean conductance $<G>$ of lattices of size b is calculated at the percolation threshold. The extrapolated y intercept is t/ν.

5. A breakdown problem in two dimensions. The algorithm described in section 3 can be used for any property for which series, parallel, and $Y-\nabla$ transformations exist. Transformations for a simple model of the supercurrent have been developed [7,8]. These calculate the maximum allowed flow through a network given the maximum flow allowed in each bond. This model is a simplification when applied to real superconductors, where phase coherence of the superconducting wavefunctions is important, but the model is still useful as a first approximation.

For two superconductors in series, the weakest of the two goes into the normal state first. Thus, for the series configuration [illustrated for conductors in Figure 1(A)],

$$i_S = min\ (i_1, i_2) \tag{9}$$

where i is the critical current of a bond. If the superconductors are in parallel, a simple model is that the critical currents of the bonds add,

$$i_P = i_1 + i_2 \quad . \tag{10}$$

For the $\nabla \rightarrow Y$ transformation, it has been shown that [8]

$$i_1 = i_A + i_B \tag{11}$$

plus cyclic permutations. The $\nabla \rightarrow Y$ transformation is slightly more complex, but can be done, and is discussed elsewhere [8].

In an infinite sample, the critical current density J_C--in two dimensions, the critical current per unit width--goes to zero as p approaces p_c from above,

$$J_C \sim (p - p_c)^{\,v} \tag{12}$$

which defines the critical exponent v. A finite-size scaling analysis of data for the critical current gave $v = v$ to within about 1% [7,8].

6. Percolative conduction in three dimensions. The node-removal transformation, shown in Figure 1(D), can, in principle, be used to solve any conductor network: One repeatedly removes nodes until only a single bond remains. This method was introduced by Fogelholm [3], and used by him to study two-dimensional lattices near the percolation threshold.

This method is being used to study percolation in higher dimensions [9]; a preliminary report of results on three-dimensional percolative lattices will be presented here.

The most efficient way to apply this method is to remove the least-interconnected nodes first, and to reduce parallel connections as soon as they occur. This tends to minimize the effects of a disadvantage of the node-removal transformation: It replaces the n bonds which were connected to the removed node with $n(n-1)/2$ new bonds. Thus, for lattices which are nearly full, the Fogelholm method is not faster than more conventional methods [1]. Near the percolation threshold, however, the method is as fast as the two-dimensional method described above: the time to solve a lattice is roughly proportional to the number of sites in the lattice in all dimensions. A $b = 80$ cubic lattice near p_c can be solved in 75 seconds on a Sun-4 computer.

Figure 5 shows the results of simulations on cubic lattices from $b = 4$ through $b = 80$. Preliminary analysis of this and other data gives $t/v - 1 \cong 1.27$. Using the three-dimensional value $v \cong 0.88$, this gives $t \cong 2.00$ in three dimensions.

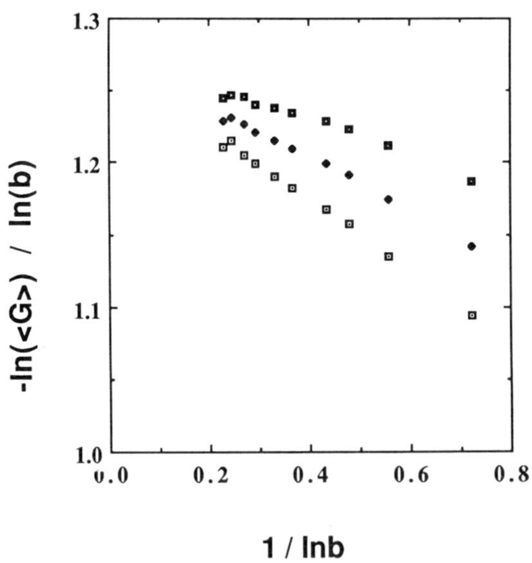

Figure 5. Plot of data from cubic lattices. The mean conductance $<G>$ of lattices of size b is calculated at the percolation threshold. An extrapolated y intercept of this data gives $t/v - 1$.

7. Discussion. This paper has described two algorithms which can be used to solve a number of circuit problems. These algorithms work by repeatedly transforming the circuit until only a single element remains which characterizes the properties of the entire circuit. For example, they can be used to find the conductance or critical current between two chosen points of a complex circuit. The methods are numerically exact (in that it does not depend on any successive approximation scheme), and are fast enough to be valuable even on small computers.

Both the two-dimensional and Fogelholm methods work with arbitrary complex impedences. The two-dimensional method can also be applied to a variety of breakdown problems, as described above, and can also be used for solving latices of Ising spins, where the necessary Y-∇ transformations exist.

Acknowledgements. I would like to thank D. J. Frank, M. Octavio, and D. Gingold for their essential contributions to the work described here.

REFERENCES

1. D. J. Frank and C. J. Lobb, *Highly efficient algorithm for percolative transport studies in two dimensions*, Phys. Rev. B **37**, 302 (1988).

2. C. J. Lobb and D. J. Frank, *Percolative conduction and the Alexander-Orbach conjecture in two dimensions,* Phys. Rev. B **30**, 4090 (1984).

3. Rabbe Fogelholm, *The conductivity of large percolation network samples,* J. Phys. C **13**, L571 (1980).

4. Stephen B. Senturia and Bruce D. Wedlock, *Electronic Circuits and Applications*, John Wiley and Sons, N.Y, 1975, p. 75.

5. M. P. M. den Nijs, *A relation between the temperature exponents of the eight-vertex model and the q-state Potts model,* J. Phys. A **12**, 1857 (1979), B. Derrida and L. de Seze, *Application of the phenomological renormalization to percolation and lattice animals in dimension 2,* J. Phys. (Paris) **43,** 475 (1982), C. J. Lobb and Keith R. Karasek, *Critical exponents for two-dimensional bond percolation,* Phys. Rev. B **25,** 492 (1982).

6. Peter J. Reynolds, H. Eugene Stanley, and W. Klein, *Large-cell Monte-Carlo renormalization group for percolation,* Phys. Rev. B **21**, 1223 (1980).

7. C. J. Lobb and D. J. Frank, *Percolation Critical Exponents for Conductance and Critical Current in Two Dimensions,* in Inhomogeneous Superconductors, D. U. Gubser, T. L. Francavilla, J. R. Leibowitz, and S. A. Wolf, ed., AIP Conf. Proc. **58,** AIP, New York, 1980, p. 308.

8. M. Octavio, A. Octavio, J. Aponte, R. Medina, and C. J. Lobb, *Nonuniversal critical behavior in the critical current of superconducting composites*, Phys. Rev. B **37**, 9292 (1988).

9. D. Gingold and C. J. Lobb, in preparation.

CHAPTER 11

Transport and Dispersion in Random Media

Sidney Redner*

Abstract. Transport and dispersion phenomena of dynamically-neutral tracer in flow through random media are discussed. For purely convective transport, a deterministic self-similar model is introduced for which the distribution of transit times for the tracer to traverse the network can be calculated. As a function of detailed geometric features of the system, the transit time distribution may be governed by a single time scale, or there may be an infinite sequence of dynamical transitions, in which progressively more time scales are needed to characterize the distribution. To treat the combined effects of convection and molecular diffusion on the transport of tracer, we study a "hydrodynamically-biased" continuum random walk on percolation clusters. A calculational method is outlined which gives the exact moments of the transit time distribution for a given network configuration. For $L \times L$ networks at the bond percolation threshold, the kth moment of the transit time, $\langle t^k \rangle$, at zero flow rate are governed only by a single diffusional time scale, $\tau_D \sim L^{k(2+\theta)}$, where θ is the anomalous diffusion exponent. At high flow rates, $\langle t^k \rangle$ scales as $\tau_c \tau_D^{k-1}$, where $\tau_C \sim L^{d_f}$ is a convective time scale. This two-time-scale behavior is explained on the basis of a simple heuristic argument and from detailed analytical calculations. We also consider dispersion in macroscopically heterogeneous media. For simple geometries, we illustrate the subtleties involved in determining the rules for composing homogeneous, but dissimilar elements. We also discuss some intriguing aspects of dispersion in layered media in the presence of random velocity fields, where superdiffusive behavior arises.

1. Introduction. Hydrodynamic dispersion is a fundamental transport process in which dynamically-neutral tracer disperses in a flow field under the combined action of convection and molecular diffusion.[1-11] The subtle interplay between these two mechanisms gives rise to a transport phenomenon which is considerably more complicated, but richer phenomenologically, than either simple diffusion or biased diffusion. A crucial feature is that dispersion can be dominated by the small fraction of tracer that is caught in relatively rare, stagnant regions within the flow field. Our goal, in this paper, is to provide a simple discussion of some of the intriguing features of dispersion and transport phenomena in random media.

Given the inherent difficulties in accounting for both convection and molecular diffusion in dispersion phenomena, it is worthwhile to consider simpler situations which are tractable analytically.[9] We therefore first consider dispersion in a self-similar hierarchical model

* Center for Polymer Studies and Department of Physics, Boston University, Boston, MA 02215. The Center for Polymer Studies is supported in part by grants from the ARO, NSF, and ONR.

in the absence of molecular diffusion. A general correspondence between the flow of fluid when each bond of the system is considered to be a one-dimensional tube, and the flow of electrical current when the bonds are considered to be identical resistors is described. This connection will be exploited repeatedly. For the hierarchical model, we introduce an adjustable "asymmetry" parameter which controls the width of the distribution of transit times for tracer to pass through the multiplicity of fluid paths through the network. We then formally solve for the distribution of transit times, and obtain explicit expressions for the moments of the transit time.

As the asymmetry of the hierarchical model is varied, the width of the transit time distribution also varies correspondingly, leading to two classes of behavior. On the other hand, for a sufficiently narrow distribution, $\langle T_N^k \rangle \sim c_k \langle T_N \rangle^k$, where $\langle T_N^k \rangle$ is the kth moment of the transit time distribution on the Nth-order hierarchy, and c_k is a numerical coefficient. This implies that a localized pulse of tracer spreads out at the same rate at which the pulse convects downstream. For a sufficiently broad distribution, $\langle T_N^k \rangle / \langle T_N \rangle^k \to \infty$ as $N \to \infty$, leading to enhanced dispersion which is dominated by the small fraction of tracer that passes through the the slowest bond in the network.

We next consider dispersion in low-porosity media where both of convection *and* diffusion drive the transport of tracer.[11] This type of medium is modeled by a percolation model, in which a lattice of identical tubes are either occupied with probability p, or absent with probability $1-p$. The analogy between electrical current and fluid flow is used to develop a random resistor network description of fluid transport and dispersion. In the presence of a steady-state background flow imposed by an external pressure drop, we calculate the detailed microscopic rules which govern the motion of the tracer at the level of single tubes and tube junctions. These rules are based on describing the combined convective and diffusive motion of the tracer in a tube in terms of a convection-diffusion equation (CDE), with the magnitude of the convection is proportional to the flow velocity in the bond. This gives rise to a "hydrodynamically-biased" continuum random walk as a model for tracer motion; this accounts for the finite macroscopic flow in a physically meaningful way. Together with the assumption of perfect mixing of fluid at tube junctions, we are able to construct a calculational approach which yields, in principle, the *exact* values for moments of transit times for tracer to traverse an arbitrary network.

For percolating networks at the percolation threshold on $L \times L$ square lattices, our numerical results suggest that at zero flow velocity, the transit time moments are all characterized by a single diffusive time scale, i.e., $\langle t^k \rangle \sim \langle t \rangle^k$. However, at high flow rates, *two* characteristic times, a convective and a diffusive time scale are needed to account for the behavior of the moments. This two-time-scale hypothesis is justified by a simple physical argument.

Finally, we consider a number of striking aspects of dispersion in macroscopically heterogeneous media. We seek to develop the composition rules for transport laws when homogeneous, but distinct subunits are superposed. For relatively simple geometries, this appears to be a non-trivial task. Moreover, when macroscopic geometrical inhomogeneity *and* inhomogeneity in velocity fields occur together, *super*diffusive behavior can occur in which the mean-square displacement grows faster than linearly in time. We discuss some of the basic features of this new type of transport phenomenon.

2. Transport and dispersion on the asymmetric hierarchical model. Consider the flow of fluid on the hierarchical model shown in Fig. 1. The system is a hierarchy of both singly-connected bonds and blobs; this appears to capture many of the essential geometrical features of the percolating backbone at the percolation threshold. An important aspect of this asymmetric hierarchy is that the curved bonds at a given level have a conductance decreased by a factor ϵ compared to the straight bonds at the same level.

We are interested in the properties of fluid flow and dispersion of tracer through this system when each bond is considered to be a one-dimensional tube. To this end, we now outline a general analogy between Poiseuille flow in the tube network and electrical current flow in a corresponding resistor network. For a resistor with conductance G, we choose the radius r and the length ℓ of the corresponding tube by $r^2 \sim G^{1/3}$ and $\ell \sim G^{-1/3}$. This gives a flow conductance, $g \sim r^4/\ell$, varying as G. In addition, when a unit pressure drop is imposed across the bond, the average fluid velocity $v \sim g/r^2$ varies as $G^{2/3}$, and the transit time $\tau \sim \ell/v$ therefore varies as G^{-1}. Thus the bond transit time is strictly proportional to the

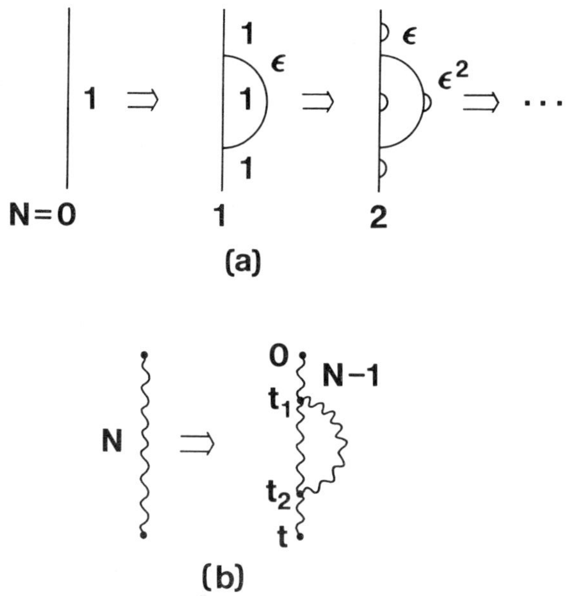

Figure 1. (a) The self-similar hierarchical model, in which a bond of conductance 1 at level N is replaced at level $N+1$ by four new bonds with conductances as indicated in the figure. Shown are the hierarchies at the $N = 0$, 1, and 2 levels. In (b) the corresponding decomposition of an Nth-order structure into four $(N-1)^{\text{th}}$-order structures. The labeling of the times used in the integral recurrence relation (2a) is indicated.

inverse of the flux in the bond. Since each bond has the same volume, the tracer spends an equal amount of time, on average, in each bond of the network. This parameterization also ensures that, in the convective limit, tracer enters a particular bond with a probability proportional to the flux in the bond.

The "asymmetry" parameter ϵ controls the width of the distribution of transit times for tracer particles to convect across the network. When $\epsilon = 1$ (symmetric limit), both paths of the first-order structure have the same transit time and there is no dispersion, while for $\epsilon \neq 1$, the variety of distinct paths available for the tracer causes a delta function input of tracer to evolve into a broadened distribution. Note that as $\epsilon \to 0$, one fluid path will have a diverging transit time while the other transit time remains finite. Consequently, the limit $\epsilon \to 0$ is a singular perturbation as flow in the slowest bond dominates in dispersion phenomena.

Let us now calculate the average time required for tracer to traverse the Nth-order structure, $\langle T_N \rangle$, when a unit potential difference (or pressure drop) is imposed across the two endpoints of the model. By decomposing the Nth-order structure into the four $(N-1)^{\text{th}}$ order structures as indicated in Fig. 1, the transit times on each of the lower-order structures will be rescaled by the factors $a \equiv (3+2\epsilon)/(1+\epsilon)$, $b \equiv 3+2\epsilon$, a, and $c \equiv (3+2\epsilon)/\epsilon$, respectively. Since the tracer takes the straight path with probability $p_{st} = 1/(1+\epsilon)$, and the curved path with probability $p_{cu} = \epsilon/(1+\epsilon)$, we find the simple recursion relation

$$\langle T_N \rangle = p_{st}(2a+b)\langle T_{N-1} \rangle + p_{cu}(2a+c)\langle T_{N-1} \rangle \equiv a_1 \langle T_{N-1} \rangle, \tag{1a}$$

with $a_1 = 4a$. Using $\langle T_0 \rangle = 1$, we thus obtain

$$\langle T_N \rangle = \langle T_1 \rangle^N = (a_1)^N. \tag{1b}$$

To compute the distribution of transit times, denote by $P_N(V;t)$ the probability density for traversing an Nth-order hierarchy in a time t when a potential drop V is applied across the system. This distribution satisfies the integral recurrence relation

$$\begin{aligned} P_N(V;t) = \int_0^{t_2} dt_1 \int_0^t dt_2 \Big[& p_{st} P_{N-1}\big(\frac{V}{a};t_1\big) P_{N-1}\big(\frac{V}{b};t_2-t_1\big) P_{N-1}\big(\frac{V}{a};t-t_2\big) \\ & + p_{cu} P_{N-1}\big(\frac{V}{a};t_1\big) P_{N-1}\big(\frac{V}{c};t_2-t_1\big) P_{N-1}\big(\frac{V}{a};t-t_2\big)\Big]. \end{aligned} \tag{2a}$$

In terms of the Laplace transform, $\tilde{P}_N(V;z) = \int P_N(V;t)e^{-zt}dt$, we obtain the functional recursion relation

$$\tilde{P}_N(z) = \left[\tilde{P}_{N-1}(az)\right]^2 \left(p_{st}\tilde{P}_{N-1}(bz) + p_{cu}\tilde{P}_{N-1}(cz)\right), \tag{2b}$$

where the argument V has been dropped since it can be rescaled to an identical value for all the P_N's.

On the other hand, the Laplace transform is the moment generating function

$$\tilde{P}_N(z) = \langle e^{-zt} \rangle = 1 - z\langle T_N \rangle + \frac{z^2}{2}\langle T_N^2 \rangle - \dots \tag{3}$$

so that by expanding (2b) in a power series, one obtains recursion relation in which the kth moment on the Nth-order structure, $\langle T_N^k \rangle$, can be decomposed in terms of all moments up to order k on the $(N-1)^{\text{th}}$-order structure. For example, for the case $k = 2$, we obtain

$$\langle T_N^2 \rangle = a_2 \langle T_{N-1}^2 \rangle + b_2 \langle T_{N-1} \rangle^2, \tag{4a}$$

with

$$a_2 = \left(\frac{1+4\epsilon+\epsilon^2}{\epsilon}\right)\left(\frac{3+2\epsilon}{1+\epsilon}\right)^2; \qquad b_2 = 10\left(\frac{3+2\epsilon}{1+\epsilon}\right)^2. \tag{4b}$$

If we iterate Eq. (4a), we ultimately arrive at

$$\langle T_N^2 \rangle = (1-c_2)a_2^N + c_2(a_1^2)^N, \tag{5}$$

with $c_2 = b_2/(a_1^2 - a_2) = 10\epsilon/(12\epsilon - \epsilon^2 - 1)$.

When $a_1^2 > a_2$, which occurs when $\epsilon > 0.08392\ldots \equiv \epsilon_2$, the second term in (5) dominates and we have, as $N \to \infty$

$$\langle T_N^2 \rangle \sim c_2(a_1^2)^N = c_2\langle T_N \rangle^2. \tag{6}$$

(Note that this gives $\langle T_N^2 \rangle = \langle T_N \rangle^2$ when $\epsilon = 1$.) Thus for $\epsilon > \epsilon_2$, both $\langle T_N \rangle$ and $\sqrt{\langle T_N^2 \rangle}$ are governed by a single characteristic time as $N \to \infty$. However for $\epsilon \le \epsilon_2$, the first term in (5) dominates, so that as $N \to \infty$

$$\langle T_N^2 \rangle \sim (1-c_2)a_2^N. \tag{7}$$

This limiting behavior for the second (and higher) moments for large N as $\epsilon \to 0$ stems from the small fraction of tracer which enters the slowest bond of the network. That is, taking the square of the time required to traverse the slowest bond in the network, and multiplying by the probability of entering this bond, gives a contribution to the second moment that equals $(9/\epsilon)^N$ for $\epsilon \to 0$, and this coincides with (7) in the $\epsilon \to 0$ limit. Thus the second moment is indeed dominated by the tracer which passes through the slowest bond in the network. Moreover, as $\epsilon \to 0$, $\langle T_N \rangle \sim 12^N$, so that the ratio $\langle T_N^2 \rangle / \langle T_N \rangle^2$ diverges as $(16\epsilon)^{-N}$. Consequently there is more than one time scale that characterizes the transit time distribution.

Similar, but more tedious, considerations apply for the higher moments. The general picture that emerges is that there exists a series of dynamical transitions as a function of ϵ in which the scaling behavior of the moments change. That is, there is an increasing sequence $\{\epsilon_k\}$, with $\epsilon_2 = 0.08392\ldots$, $\epsilon_3 = 0.14727\ldots$, $\epsilon_4 = 0.188003\ldots$, with $\lim_{k\to\infty} \epsilon_k = 1/3$, such that for $\epsilon_k \leq \epsilon < \epsilon_{k+1}$ moments of order $j < k$, are governed by a single time scale, namely

$$\langle T_N^j \rangle \sim c_j (a_1^j)^N = c_j \langle T_N \rangle^j, \tag{8a}$$

while moments of order $j \geq k$, scale anomalously, *i.e.*,

$$\langle T_N^j \rangle \sim z_j a_j^N. \tag{8b}$$

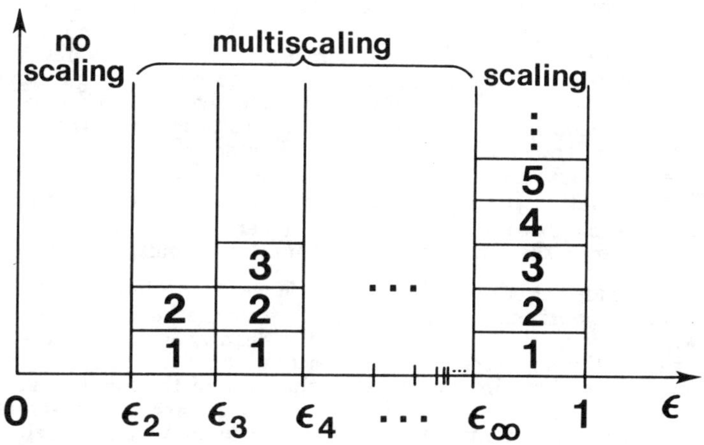

Figure 2. "Phase diagram" for the behavior of the transit time moments of the hierarchical model as a function of the asymmetry parameter ϵ. The labelled blocks indicate the orders of the transit time moments which obey one-parameter scaling.

In general, for $\epsilon > 1/3$, the relation $\langle T_N^k \rangle \sim c_k \langle T_N \rangle^k$ holds for all k. That is, a single time scale, $\langle T_N \rangle$, characterizes *all* the moments of the transit time distribution. We term this behavior in the weakly asymmetric limit as the "scaling" regime. On the other hand for $\epsilon < \epsilon_2$, each moment scales independently. This behavior arises solely from the small fraction of tracer which enters the slowest bond in the network. In the limit $\epsilon \to 0$, this fraction gives the following contribution to the kth-moment

$$\left(\epsilon \cdot \left(\frac{3}{\epsilon}\right)^k\right)^N, \tag{9}$$

which coincides with the limit $\epsilon \to 0$ behavior of a_k^N. We term this new type behavior as "multiscaling".

In the scaling regime, the functional recursion relation for the Laplace transform of the transit time distribution can be simplified by using $\langle T_N^k \rangle \sim c_k \langle T_N \rangle^k$ in (3) to yield, for large N

$$\tilde{P}_N(z) = \langle e^{-zt} \rangle = 1 - z\langle T_N \rangle + c_2 \frac{z^2}{2} \langle T_N \rangle^2 - \ldots \equiv \tilde{g}(z\langle T_N \rangle). \tag{10}$$

That is, $\tilde{P}_N(z)$ is a function only of the scaling combination $z\langle T_N\rangle = z(4a)^N$. Using (10) in (2b), we find the following functional relation for the scaling function $\tilde{g}(z)$

$$\tilde{g}(z) = \left[\tilde{g}\left(\frac{z}{4}\right)\right]^2 \left(p_{st}\tilde{g}(\frac{bz}{4a}) + p_{cu}\tilde{g}(\frac{cz}{4a})\right), \tag{11}$$

with asymptotic solution as $z \to \infty$

$$\tilde{g}(z) \sim A\exp(-(z/z_0)^\alpha), \tag{12}$$

where $A = \sqrt{1+\epsilon}$, z_0 is arbitrary, and α is determined by the root of $4^\alpha = 2 + (1+\epsilon)^\alpha$. When $\epsilon = 1$, this gives $\alpha = 1$, which corresponds to a delta function for the transit time distribution. On the other hand, for $\epsilon < 1$, α is also less than 1, and this quasi-exponential decay in z implies a broadened distribution of transit times.

3. Dispersion in the presence of percolation disorder. For a poorly connected medium, percolation offers a simple and accurate description of many transport properties.[12] We will again exploit the analogy between fluid flow in a random network of tubes and electrical current flow in the corresponding random resistor network to develop a formalism to compute the distribution of transit times for tracer to pass through a percolating network. Our approach is based on first solving for the steady-state fluid flow in the network when a fixed pressure drop is imposed across the opposite edges of the system, and then solving for the motion of tracer through the network under the action of the imposed flow. The detailed rules for tracer motion are based on a picture in which the tracer obeys the convection-diffusion equation (CDE) in each tube with the magnitude of the convection is proportional to the local pressure drop in the bond, and on the assumption that there is perfect mixing of tracer at the junctions between multiple bonds.

Thus we are considering a *hydrodynamically-biased, continuum* random walk with an inhomogeneous bias proportional to the current flow in the bond on the backbone of the network, and with isotropic diffusive motion on the dead ends where there is no flow. We are considering the continuum description in order to account for the longitudinal dispersion within each time, *i.e.*, we have implicitly taken the ratio of the bond length to the mean-free path to be infinite. This feature should provide a more accurate description of motion in a real random network, as compared to a discrete random walk model. An additional noteworthy point is the contrast between the hydrodynamically-biased and *globally*-biased random walk models.[13,14] In the latter model, it is typically assumed that there is a uniform bias everywhere, *including* the dead ends. This leads to dead end residence times which grows exponentially with the length of the dead end ℓ, rather than growing as ℓ^2 if there were unbiased diffusion. Consequently, trapping in the dead ends plays an anomalously large role in determining the motion of a random walker. This might be appropriate for an extremely dilute system where hydrodynamics is not relevant, but not for typical fluid flow.

In order to solve the problem of tracer motion through a percolating network, we first require the solution to the CDE for a single one-dimensional tube.[11] The CDE is

$$\frac{\partial c(x,t)}{\partial t} + u\frac{\partial c(x,t)}{\partial x} - D_{\rm mol}\frac{\partial^2 c(x,t)}{\partial x^2} = 0, \tag{13}$$

where $c(x,t)$ is the tracer concentration at position x and time t, u is the flow velocity, and $D_{\rm mol}$ is the molecular diffusion coefficient. We require the solution to the CDE with boundary conditions appropriate for injection of tracer at one end of the tube, and removal of the tracer at the other end. This corresponds to a unit flux at the inlet end, $j(x = 0, t) = \delta(t)$, and zero concentration at the outlet end, $c(x, t = 0) = 0$. Notice that the flux $j(x,t) = uc(x,t) - D_{\rm mol}\frac{\partial c(x,t)}{\partial x}$ has a contributions arising from both convection and dispersion.

For an initially empty system of length L, the CDE is conveniently solved by introducing the Laplace transform, $\tilde{c}(x,s) = \int_0^\infty c(x,t)e^{-st}dt$ to yield a solution of the form

$$\tilde{c}(x,s) = Ae^{\alpha x} + Be^{\beta x}, \tag{14}$$

where $\alpha, \beta = (u \pm \sqrt{u^2 + 4D_{\text{mol}}s})/2D_{\text{mol}}$, and from the boundary conditions, we determine the constants to be $A = [D_{\text{mol}}(\beta - \alpha e^{(\alpha-\beta)L})]^{-1}$ and $B = [D_{\text{mol}}(\alpha - \beta e^{-(\alpha-\beta)L})]^{-1}$. For a unit flux input and an absorbing boundary at the outlet, we have the fundamental correspondence that the first passage probability density, $P(t)$, coincides with the outlet flux. Thus the Laplace transform of $P(t)$ is

$$\tilde{P}(s) = \tilde{j}(x = L, s) = \frac{m_s e^m}{m \sinh m_s + m_s \cosh m_s}, \tag{15}$$

where $m = uL/2D_{\text{mol}}$ is the Peclet number, and where $m_s = \sqrt{m^2 + sL^2/D_{\text{mol}}}$.

Since $\tilde{P}(s)$ is the generating function of the transit time moments, they can be extracted from the series expansion of $\tilde{P}(s)$ in powers of s. We thereby obtain

$$\langle t \rangle = \frac{L^2}{D_{\text{mol}}}(\frac{1}{2m} - \frac{1}{4m^2}[1 - e^{-2m}]), \tag{16a}$$

$$\langle t^2 \rangle = (\frac{L^2}{D_{\text{mol}}})^2(\frac{1}{4m^2} - \frac{1}{8m^4}[2 - (6m+1)e^{-2m} - e^{-4m}]), \tag{16b}$$

and so on. The limiting forms of the moments are instructive: For the first moment, we find

$$\langle t \rangle \to \frac{L^2}{2D_{\text{mol}}} \qquad m \to 0, \tag{17a}$$

$$\to \frac{L}{U} - \frac{D_{\text{mol}}}{U^2} \qquad m \to \infty, \tag{17b}$$

while for the second moment,

$$\langle t^2 \rangle \to \frac{5}{12}\left(\frac{L^2}{D_{\text{mol}}}\right)^2 \qquad m \to 0, \tag{18a}$$

$$\to \frac{L^2}{U^2} - \frac{2D^2}{U^4} + \mathcal{O}(e^{-m}) \qquad m \to \infty, \tag{18b}$$

and the limiting behavior for an arbitrary moment has the form

$$\langle t^k \rangle \to x_k\left(\frac{L^2}{D_{\text{mol}}}\right)^k \qquad m \to 0, \tag{19a}$$

$$\to \frac{L^k}{U^k} + \ldots \qquad m \to \infty, \tag{19b}$$

where the x_k are constants. Thus at small Peclet numbers, the transit time moments are all governed by the diffusion time, L^2/D_{mol}, while at high Peclet numbers the moments are governed by the convection time, L/U.

We now extend this calculational approach to treat a network of nodes i connected by one-dimensional tubes ij of length l_{ij} and cross-sectional area S_{ij}. For simplicity we henceforth take the l_{ij} to have a common value l; however, the general case can be treated straightforwardly. The nodes at one end of the network are all connected to an inlet node I, and those at the other end to an outlet node O. We first compute the background flow field by applying a fixed pressure drop between the inlet and outlet nodes. Due to the complete mixing of the tracer at the nodes of the network, each tube "sees" the rest of the system only through the tracer concentration at its two ends. Thus, as we shall describe, the tracer motion in the network can be accounted for in terms of a (non-symmetric) matrix equation for the node concentrations.

As in the single tube problem, we require the first passage probability density, $P(t)$, for the tracer to arrive at the outlet node, given that there is a unit delta function injection of flux at the inlet node. This first passage problem may be solved in terms of the set of tracer concentrations, $\{c_i(t)\}$, at the nodes of the network. To solve for the $\{c_i(t)\}$, first consider tube ij in which the tracer concentration within the tube, $c_{ij}(x,t)$, satisfies the one-dimensional CDE, where the convection term u now varies in space according to the magnitude of the background flow, *i.e.*, $u = u_{ij}$. We now solve this system of equations for an initially empty system, subject to the boundary conditions appropriate for continuity of the tracer concentrations at all junctions. That is, $\tilde{c}_{ij}(x,s) = \tilde{c}_i$ at $x = 0$, and $\tilde{c}_{ij} = \tilde{c}_j$ at $x = l$. In close analogy with the case of a single tube, the solution is of the form $\tilde{c}_{ij}(x,s) = A_{ij}e^{\alpha_{ij}x} + B_{ij}e^{\beta_{ij}x}$, with $\alpha_{ij}, \beta_{ij} = (u_{ij} \pm \sqrt{u_{ij}^2 + 4D_{\mathrm{mol}}s})/2D_{\mathrm{mol}}$, and

$$A_{ij} = \frac{\tilde{c}_j - \tilde{c}_i e^{\beta_{ij}l}}{e^{\alpha_{ij}l} - e^{\beta_{ij}l}} \qquad B_{ij} = \frac{\tilde{c}_i e^{\alpha_{ij}l} - \tilde{c}_j}{e^{\alpha_{ij}l} - e^{\beta_{ij}l}}. \tag{20}$$

From this, the Laplace transform of the flux leaving node i along tube ij is

$$\tilde{j}_{ij} = S_{ij}\left(u_{ij}\tilde{c}_{ij} - D_{\mathrm{mol}}\frac{\partial \tilde{c}_{ij}}{\partial x}\right)_{x=l}. \tag{21}$$

We may now express this flux in terms of the concentrations $\tilde{c}_i$ and $\tilde{c}_j$ at the two ends of the tube. Thus writing $\tilde{J}_{ij} = G_{ij}^{+}\tilde{c}_i - G_{ij}^{-}\tilde{c}_j$, we have

$$G_{ij}^{+}(s) = D_{\mathrm{mol}}S_{ij}(m + m_s \coth m_s)/l, \tag{22a}$$
$$G_{ij}^{-}(s) = D_{\mathrm{mol}}S_{ij}\left(m_s e^{-m}/\sinh m_s\right)/l. \tag{22b}$$

For simplicity we have dropped in subscript ij on m and m_s. For nodes of negligible volume with no accumulation of tracer possible, then the $\tilde{c}_i$ satisfy the current conservation conditions (for the interior nodes of the network),

$$\sum_j \tilde{j}_{ij} = \sum_j (G_{ij}^{+}\tilde{c}_i - G_{ij}^{-}\tilde{c}_j) = 0, \tag{23a}$$

while at the inlet node,

$$\sum_j \tilde{J}_{Ij} = \sum_j (G_{Ij}^{+}\tilde{c}_I - G_{Ij}^{-}\tilde{c}_j) = 1, \tag{23b}$$

corresponding to a delta function input of flux. Due to the absorbing boundary condition at the outlet node, the Laplace transform of the first passage probability density, $P(t)$, again coincides with the flux exiting via the outlet node

$$\tilde{P}(s) = -\sum_j \tilde{J}_{Oj} = \sum_j G_{Oj}^{-}\tilde{c}_j. \tag{24}$$

For any value of s, Eqs. (23) and (24) are the analogues of the Kirchhoff equations, $\sum_j g_{ij}(v_i - v_j) = 0$, for the potentials v_i in an electrical network. The primary difference between the two cases is that in the tracer problem, the effective tube conductances are direction dependent, leading to non-symmetric network equations.

Eqs. (23) may be solved in a number of ways. In [15], they were solved by a "probability propagation" algorithm which was motivated by a physical picture in which individual packets of first passage probability are propagated along all possible paths through the system. By this technique, the network first-passage probability $\tilde{P}(s)$ can be obtained exactly, in principle, up to geometric convergence inherent in summing over progressively longer and longer paths. For a well-connected network with little backflow or stagnation effects, probability propagation leads to an efficient method as the computation time is proportional to the system volume. Consequently, this method is intrinsically superior to a computational approach which propagates individual particles through the network. In fact, a finite simulation based on such an approach must lead to incorrect results, as the low-probability paths, which can dominate the higher transit-time moments, will not be sampled. For a poorly-connected medium, however, backflow and stagnation cause the propagation of first passage probability to become relatively inefficient. In this case, it is more efficient to solve the network equations (23) directly.[10,11]

Our simulation results are based on solving square lattice networks of identical tubes at the percolation threshold, with the linear dimension L ranging from 2 to 70. For each configuration, we first find the background flow field by solving for the node potentials of the corresponding random resistor network with a unit external potential drop applied. Then we solve for the tracer motion according to the calculational scheme outlined above. The primary results of the simulation are the following. In the limit of zero flow rate (no convection) the tracer motion reduces to a continuum version of anomalous diffusion[16] on percolation clusters. Owing to the self-similar geometry, one expects that the diffusion equation does not apply, but rather, that the transport is described by anomalous diffusion in which

$$\langle t^k \rangle \sim (\tau_D)^k \sim L^{k(2+\theta)}. \tag{25}$$

Our data are consistent with the previously-estimated value $2+\theta \approx 2.86$ in two dimensions.[14]

The behavior in the high Peclet number regime is considerably more interesting. First, the reduced moment $\langle t^k \rangle^{1/k}$ now *depends* on k, indicating that there is no unique time scale which characterizes the transit time distribution. However, it is possible to give a very simple argument which appears to account for this unusual scaling of the moments. Consider the Y-shaped network of three tubes, each of volume v (Fig. 3), where the incoming fluid flux q branches into two tubes of respective fluxes ϵq and $(1-\epsilon)q$. It is pedagogically useful to think of the "slower" branch as approaching a dead end as the limit $\epsilon \to 0$ is taken. The tracer flux through each branch may be viewed as the sum of diffusive and convective processes which occur independently and simultaneously. The total transition rate into the slower branch is then the sum of its diffusive rate, equal to the inverse diffusion time $D_{\rm mol}/l^2$, and its convective rate, equal to the inverse convection time $u/l = \epsilon q/v$, where u is the velocity in the upper branch. For the "faster" branch, the transition rate is dominated by the convective rate, q/v. As $\epsilon \to 0$, the transition rate into the slower branch is controlled by diffusion and therefore tends to $D_{\rm mol}/l^2$. Correspondingly, the relative entrance probabilities tend to $p_u = D_{\rm mol}v/l^2 q + \mathcal{O}(q^{-2})$ and $p_l = 1-p_u$. Thus the moments of the transit time distribution are, to leading order in q^{-1},

$$\langle t^k \rangle = p_f t_f^k + p_s t_s^k \simeq \frac{D_{\rm mol} v}{l^2 q}\left(\frac{v}{q} + \frac{l^2}{D_{\rm mol}} + \frac{v}{q}\right)^k + 1 \cdot \left(\frac{v}{q} + \frac{v}{q}\right)^k, \tag{26}$$

where the subscript f and s refer to the faster and slower branches, respectively. We rewrite this result in the simpler (and suggestive) form

$$\langle t^k \rangle \sim \tau_C \tau_D^{k-1}, \tag{27}$$

where $\tau_C = \langle t \rangle = 3v/q$, is the "convection" time, which, in turn, is equal to the total volume of the system divided by the flux, and $\tau_D = l^2/D_{\rm mol}$ is the "diffusion" time in the slower branch. Thus the contribution of the slowest bond dominates in all but the first moment, and for this particular case, the slowest bond contribution is of the same order as that of the rest of the network, leading to a transit time which is exactly proportional to the volume of the system.[11,17,18]

To make a correspondence with the transit time moments on percolating clusters, we identify τ_C and τ_D with percolation quantities. From the theorem that the average transit

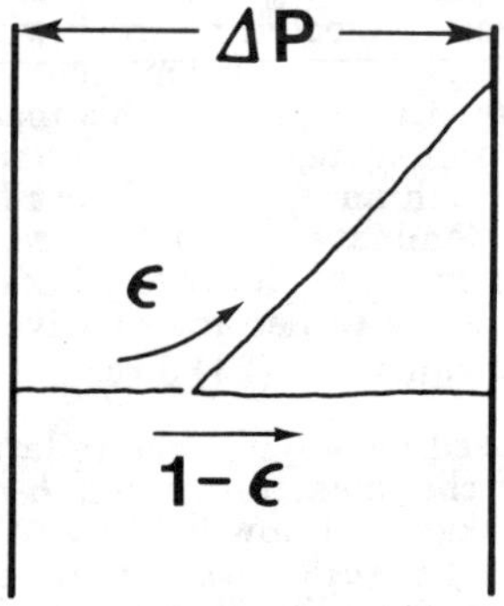

Figure 3. A simple network containing a primarily diffusive path and a primarily convective path. Each tube is of volume v.

time at high flow rates is proportional to the volume of the system, we have $\tau_C \sim L^{d_f}/I$, *i.e.*, the average number of bonds in a percolating cluster of an $L \times L$ system divided by the total current flow I. From our numerical results at zero flow rate, there appears to be a unique diffusive time scale $\tau_D \sim L^{2+\theta}$, which governs all the transit time moments in the diffusive limit. Combining these two results, we anticipate that

$$\langle t^k \rangle \sim \frac{1}{I} L^{d_f+(k-1)(2+\theta)}. \tag{28}$$

This form provides an excellent account of our numerical data.

4. Dispersion in Macroscopically Heterogeneous Media. Thus far we have considered situations systems for which the disorder was on the scale of a single bond. We now discuss some novel transport phenomena that can arise when fluid is flowing through a medium in which the disorder is correlated on a *macroscopic* scale. Consider, for example, a stratified medium in which fluid is diffusing perpendicular to the layers (Fig. 4). Let us compute the first passage time for tracer to traverse a finite number of layers. For N layers of lengths L_i, with respective diffusion coefficients D_i and D_2, we need to solve the diffusion equation with the boundary conditions of unit injection of flux at the left end of block 1, zero concentration at the far end of block N, and continuity of the concentrations and fluxes at all the interfaces. Using standard techniques, the Laplace transform of the first passage probability across a 2-layer composite is,

$$\tilde{P}(s) = [\cosh x_1 \cosh x_2 + \frac{D_1}{D_2} \sinh x_1 \sinh x_2]^{-1}, \tag{29}$$

where $x_i = \sqrt{sL_i^2/D_i}$, and s is the Laplace transform variable. (Notice that for $D_1 = D_2$, Eq. (29) reduces to the first passage probability for a homogeneous block of length $L_1 + L_2$.) The corresponding average first passage time is

$$\langle t \rangle = \frac{L_1^2}{2D_1} + \frac{L_2^2}{2D_2} + \frac{L_1 L_2}{D_2}, \tag{30}$$

which, interestingly, is *not* symmetric in the indices 1 and 2. Therefore $\langle t \rangle$ *cannot* be recast in terms of effective geometric and transport parameters, *e.g.*, as $\langle t \rangle = L_{\text{eff}}^2/D_{\text{eff}}$, which one might have expected if the composite could be described by an effective diffusion equation.

A similar calculation for the N-layer system yields, for the average first passage time through the composite

$$\langle t \rangle = \sum_{i=1}^{N} \frac{L_i^2}{D_i} + \frac{1}{2} \sum_{i<j}^{N} \frac{L_i L_j}{D_j}. \tag{31}$$

This form has a number of striking implications. Even for simple special cases, such as a periodic array of alternating blocks, it is not possible to recast this expression into a form that suggestive of an effective diffusion equation description for the composite. Roughly speaking, the essential problem is that lengths appear as quadratic products, while diffusion constants tend to combine inversely, leading to an anomalous composition rule for $\langle t \rangle$. More generally, it should prove interesting to learn how to write the composition rules and macroscopic description for both series and parallel superposition of homogeneous, but different, subsystems.

For the case of transport parallel to the layers, extremely intriguing behavior can arise, as Matheron and de Marsily[19] were apparently the first to recognize. Consider, as they did, many independent and homogeneous layers of differing thickness, in which the fluid velocity in each layer is a random variable which is drawn from a symmetric distribution with a zero mean (Fig. 4). The tracer therefore obeys the convection-diffusion equation, in which the (layer-dependent) convection term is given by the velocity in the layer. In the longitudinal direction, the spread of dynamically-neutral tracer was found to increase as $t^{3/4}$. This remarkable result arises because in a finite time, a random walker explores only a finite number of layers, N, within which the mean velocity is of order $N^{1/2}$ and fluctuating in sign.[19,20] This non-stationary average velocity leads to an anomalous time dependence of the longitudinal displacement.

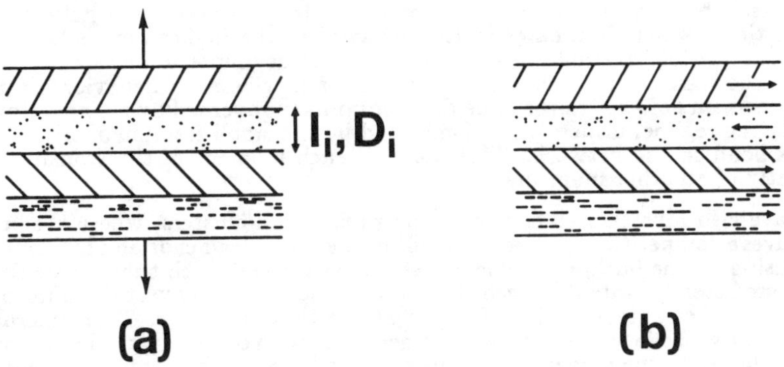

Figure 4. Schematic illustration of a stratified medium. In (a), transport perpendicular to the layers is depicted. In (b) there is a constant fluid velocity within each layer which is a random variable that is symmetrically distributed about a zero mean.

More quantitatively, assume that the mixing of fluid between different layers is driven by molecular diffusion, which is the same in all layers. Then in a time t the random walk visits $\sqrt{D_\perp t}$ layers. The average velocity within these layers is simply

$$\langle v \rangle_t = \frac{1}{\sqrt{D_\perp t}} \sum_{i=1}^{\sqrt{D_\perp t}} v_i \sim (D_\perp t)^{-1/4}, \tag{32}$$

where the subscript denotes a restricted average over only t layers. The rms longitudinal displacement at time t, is then

$$\sqrt{\langle x(t)^2 \rangle} \sim \langle v \rangle_t t \sim D_\perp^{-1/4} t^{3/4}. \tag{33}$$

This simple, but striking result raises many interesting and open questions. First, we would like to find the probability distribution for the longitudinal displacement. Our simulations suggest that this distribution can be fit to a quasi-exponential form

$$P(x,t) \propto \frac{1}{t^\nu} \exp[-(x/t^\nu)^\delta] \tag{34}$$

with $\nu = 3/4$, and with the exponent $\delta \lesssim 1.75$. This exponent value for δ is unexpected, since in most examples δ and ν satisfy the relation $\delta = (1-\nu)^{-1}$. On the other hand, a Lifshitz-type argument based on considering the walks which contribute to the extreme large-distance tail of the distribution suggests that $\delta = 4/3$. If this is indeed the case, then the form of the probability distribution is likely to be masked by very slow crossover effects, and this seems to be observed in our simulations.

Some interesting generalizations of the random walk in a stratified medium include considering d'-dimensional strata in a d-dimensional system. There exist only a very few results about the dynamics of random walks in such media. It should also prove fruitful to consider an *isotropic* version of the stratified random walk. For example consider a random walk on a random "Manhattan" grid, in which the directionality along any Avenue or Street is fixed along its entire length, but random in orientation. For this system, a simple generalization of the arguments leading to Eq. (33) suggest that the exponent ν is 2/3, and that $\delta = 3$. These types of superdiffusive behavior in systems with correlated, but disordered velocity fields should prove to be a rich area for future investigations.

5. Summary and discussion. We have treated a variety of dispersion phenomena within the framework of simple network and random walk models. For random tube networks models of disordered media, we have argued that dispersion can be dominated by the infinitesimal fraction of fluid which is stuck in the "slowest" flow regions of the system. In such a case, the distribution of times for tracer to pass through a network exhibits a tail at long times which dominates in the behavior of the higher moments of the transit time. For the specific example of dispersion on the asymmetric hierarchical model, the moments of the transit time exhibit two very different classes of behavior. In the scaling regime, all moments of the transit time distribution are governed by a single time scale. In the multiscaling regime, dispersion is dominated by the small fraction of tracer that enters the slowest bond in the network. This leads to each moment of the transit time scaling independently of all other moments.

We have also investigated dispersion on percolation clusters, where the transport of tracer involves a competition between pure diffusion in the dead ends and hydrodynamically-biased diffusion on the backbone. The tracer moves through each tube under the action of the background steady-state flow according to the rules of the convection-diffusion equation in each tube with complete mixing of tracer at the tube junctions. This general approach represents what we believe to be the correct generalization of the classical random walk on a percolation cluster to the case of finite flow rates and also to the continuum limit. We have thus developed and implemented numerically a computational recipe which yields a formally exact solution for the the transit time distribution of the tracer on a given configuration, for arbitrary values of the fluid flow velocity. Furthermore, we have postulated a scaling form for the transit time moments which is in excellent agreement with our numerical data on percolation clusters.

Finally, we have discussed the phenomenon of transport in macroscopically heterogeneous media. Even at the most elementary level of consideration, these systems exhibit a variety of novel diffusive and superdiffusive transport laws. There are still a wide variety of intriguing and puzzling questions for which satisfactory, first-principles explanations are still lacking. We believe that some of the theoretical approaches outlined in this paper will be useful in providing some clues towards understanding transport in these types of media.

6. Acknowledgments. I would like to thank J. Koplik and D. Wilkinson for a most enjoyable collaboration on dispersion phenomena. I also thank F. Leyvraz for a number of very enlightening discussions on the stratified random walk problem.

REFERENCES

1. G. DE JOSSELIN DE JONG, *Longitudinal and transverse diffusion in granular deposits*, Trans. Am. Geophys. Union, 39 (1958), pp. 67-74.

2. P. G. SAFFMAN, *A theory of dispersion in porous media*, J. Fluid Mech., 6 (1959), pp. 321-349.

3. P. G. SAFFMAN, *Dispersion due to molecular diffusion and macroscopic mixing in flow through a network of capillaries*, J. Fluid Mech., 7 (1960), pp. 194-208.

4. J. BEAR, *Dynamics of Fluids in Porous Media*, Elsevier, Amsterdam, 1971.

5. A. E. SCHEIDEGGER, *The Physics of Flow in Porous Media* University of Toronto Press, Toronto, 1974.

6. P. G. DE GENNES, *Hydrodynamic dispersion in unsaturated porous media*, J. Fluid Mech., 136 (1983), pp. 189-200.

7. M. SAHIMI, H. T. DAVIS, and L. E. SCRIVEN, *Dispersion in disordered porous media*, Chem. Eng. Commun., 23 (1983), pp. 819-831.

8. M. SAHIMI, B. D. HUGHES, L. E. SCRIVEN, and H. T. DAVIS, *Dispersion in flow through porous media—I. One-phase flow*, Chem. Eng. Sci., 41 (1986), pp. 2103-2122.

9. S. REDNER, J. KOPLIK, and D. WILKINSON, *Dispersion in a self-similar geometry*, J. Phys. A, 20 (1987), pp. 1543-1555.

10. S. ROUX, C. MITESCU, E. CHARLAIX, and C. BAUDET, *Transfer-matrix algorithm for convection-biased diffusion*, J. Phys. A, 19 (1986), pp. 65-70.

11. J. KOPLIK, S. REDNER, and D. WILKINSON, *Transport and dispersion in random networks with percolation disorder*, Phys. Rev. A, 37 (1988), pp. 2619-2636.

12. D. STAUFFER, *Introduction of percolation theory*, Taylor & Francis, London, 1985.

13. M. BARMA and D. DHAR, *Direction diffusion in a percolation network*, J. Phys. C, 16 (1983), pp. 1451-1458.

14. S. HAVLIN AND D. BEN-AVRAHAM, *Diffusion in disordered media*, Adv. Phys., 36 (1987), pp. 695-798.

15. L. DE ARCANGELIS, J. KOPLIK, S. REDNER, and D. WILKINSON, *Hydrodynamic dispersion in network models of porous media*, Phys. Rev. Lett., 57 (1986), pp. 996-999.

16. Y. GEFEN, A. ALEXANDER, and A. AHARONY, *Anomalous diffusion on percolating clusters*, Phys. Rev. Lett., 50 (1982), pp. 77-80.

17. D. B. SPALDING, *A note of mean residence time in steady flows of arbitrary complexity*, Chem. Engr. Sci., 9 (1958), pp. 74-77.

18. P. V. DANCKWERTS, *Local residence times in continuous flow systems*, Chem. Engr. Sci., 9 (1958), pp. 78-79.

19. G. MATHERON and G. DE MARSILY, *Is transport in porous media always diffusive? A counterexample*, Water Resources Res., 16 (1980), pp. 901-917.

20. A. GEORGES, *Anomalous diffusion in disordered media: statistical mechanics, theoretical models, and applications*, Ph.D. Thesis, Université de Paris-Sud, 1988.

PART III

Conference Abstracts

An Effective Continuum Theory for Estimating Transport Parameters in Two–Phase Random Media

Andreas Acrivos

The Levich Institute
City College of New York
New York, NY 10031

A new method [1] is presented for resolving the non–absolutely convergent and divergent integrals which often arise in the course of calculating the effective transport parameters of two–phase materials having random structure. Although several well-established techniques for accomplishing this have already been discussed previously in the literature [2–4], this new method appears to be more direct and to have a stronger physical basis than those formerly proposed. Briefly, its main feature is that it replaces part of the composite in the system with an effective continuum having the bulk property being calculated. By means of this step, long range interactions between particles are accounted for in a more complete way, and as a result, the convergence difficulties are eliminated.

To illustrate its use, this method is applied to three different problems which involve the calculations of:

(1) The effective thermal conductivity of a randomly distributed isotropic, and stationary suspension of identical spheres to $O(c^2)$.
(2) The sedimentation rate of a dilute random suspension of identical solid spheres settling under the influence of gravity to $O(c)$.
(3) The effective thermal conductivity and elongational viscosity of a suspension of infinitely conducting aligned slender rods, for particle concentrations ranging from infinite dilution to the semi–dilute regime.

Finally, it is pointed out that this effective continuum method is useful for estimating such transport quantities on its own without resorting to detailed many body calculations, and that it can be extended to higher orders in the volume fraction of the dispersed phase by means of a self–consistent scheme.

[1] A. Acrivos and E. Chang, *Phys. Fluids* 29, 1986, p. 1.
[2] G.K. Batchelor, *J. Fluid Mech.* 52, 1972, p. 245.
[3] D.J. Jeffrey, *Proc. Royal Soc.* A335, 1973, p. 355.
[4] R.W. O'Brien, *J. Fluid Mech.* 91, 1979, p. 17.
[5] A. Acrivos and E.S.G. Shaqfeh, *Phys. Fluids* 81, 1988, p. 1841.

Optimal Bounds and Microgeometries for Polycrystalline Aggregates

Marco Avellaneda

Courant Institute of Mathematical Sciences
251 Mercer Street
New York, NY 10012

The class of composites which can be regarded as polycrystals includes metals, compacted powders and certain solidified gases. A typical specimen is composed of an aggregate or grains or crystallites of one or more homogeneous, anisotropic phases. Due to the changes in crystal orientation, the physical properties (such as conductivity or stiffness) fluctuate throughout the material and the resulting electromagnetic or elastic fields are complex and difficult to calculate.

I will discuss several new results and techniques recently developed in collaboration with Graeme Milton, K.A. Lurie and A.V. Cherkaev, to obtain bounds on the effective properties of single–phase polycrystals (i.e. aggregates in which the crystallites are made a single given homogeneous anisotropic crystal). An example of such a material is a Z_n polycrystal in which the "single crystal" is transversely isotropic.

Our bounds are, depending on the property under consideration, consistent with either a Voigt–Reuss–type hypothesis—the corresponding field is uniform throughout the material—or with a new field optimality condition: the field is *proportional* to a uniform one, but the proportionality factor may vary continuously from one point to another. This type of field corresponds to new optimal bounds that we obtain by bounding the free energy from below by a total divergence or null–Lagrangian.

The following are more or less standard references:

[1] Z. Hashin, *J. Appl. Mech.* 50, 1983, p. 481.

[2] Z. Hashin and S. Shtrikman, *J. Mech. Phys. Solids* 10, 1962, pp. 343–352.

[3] J.P. Willis, in *Advances in Appl. Mech.* 21, Yih ed., Academic Press, New York, 1981.

[4] L.J. Walpole, in *Advances in Appl. Mech.* 21, Yih ed., Academic Press, New York, 1981.

The following references are related to the more recent work on polycrystals:

[1] M. Avellaneda, A.V. Cherkaev, K.A. Lurie, and G.W. Milton, *J. Appl. Phys.* 63, 1988, pp. 4989–5003.

[2] M. Avellaneda and G.W. Milton, *SIAM J. Appl. Math.*, 49, 1988, pp. 824–837.

[3] M. Ferrari and G. Johnson, *J. Appl. Phys.*, 63, 1988, p. 4460.

[4] K. Schulgasser, *J. Physics C* 10, 1977, pp. 407–414.

[5] K. Schulgasser, *J. Appl. Phys.* 54, 1982, pp. 1380–1382.

An Intuitive Approach and a Diagrammatic Formulation of the Renormalized Polarizability in the Maxwell–Garnett Theory

Ruben G. Barrera

Instituto de Fisica, UNAM
Apdo. Postal 20–364
01000 Mexico, D.F. Mexico

We treat the problem of calculating the effective dielectric response ϵ_M of a system of identical spheres embedded in a homogeneous material and located at random positions. First we derive a formula which relates ϵ_M with the response to an external field. Then we introduce the concept of a renormalized polarizability α^* of the spheres which takes into account, in an approximate way, the fluctuations of the local field. We derive a formula of the Clausius–Mossotti type but with the renormalized polarizability α^* instead of the bare polarizability α.

Through the use of a self–consistency condition we obtain that α^* obeys a simple second order algebraic equation with coefficients which depend on α, the volume fraction of the spheres and a functional of its two–particle distribution function. We apply our theory to a system of metallic spheres embedded in dispersionless gelatin. We show calculations of $Im\epsilon_M$ as a function of frequency and we discuss its physical interpretation. We compare our results with the ones of other theories as well as with experiment.

In the second part of this work we derive a very general expression for ϵ_M as a series expansion. One associates a graph to each term in the series and we show how to perform summations of infinite classes of graphs. Besides recovering well–known results we also obtain an explicit new formula for the renormalized polarizability α^*. We realize that our earlier result for α^* is an approximation which can be expressed in a diagrammatic language. We present numerical results for the same system as before and we compare both approaches. Finally we point out some other possibilities of infinite summations of graphs.

[1] A. Liebsch and P. Villaseñor, *Phys. Rev. B* 29, 1984, p. 6907.
[2] V.A. Davis and L. Schwartz, *Phys. Rev. B* 33, 1986, p. 6627.
[3] V.A. Davis and L. Schwartz, *Phys. Rev. B* 31, 1985, p. 5155.
[4] B.U. Felderhof and R.B. Jones, *Z. Phys. B* 62, 1986, p. 231.
[5] B.U. Felderhof and R.B. Jones, *Z. Phys. B* 62, 1986, p. 225.
[6] B.U. Felderhof and R.B. Jones, *Z. Phys. B* 62, 1986, p. 215.
[7] M.S. Wertheim, *Vol. Phys.* 25, 1973, p. 211.
[8] M.S. Wertheim, *Ann. Rev. Phys. Chem.* 30, 1979, p. 471.
[9] J.D. Ramshaw, *J. Stat. Phys.* 35, 1984, p. 49.
[10] R.G. Barrera, G. Monsivais, and W.L. Mochán, *Phys. Rev. B*, Sept. 1988.

Onset of Superfluidity in Random Media

Daniel S. Fisher

Dept. of Physics, Jadwin Hall
Princeton University, NJ 08544

In a random medium, quantum fluctuations can destroy superfluidity even at zero temperature. The disordered or "Bose glass" phase is somewhat analogous to the localized phase of non–interacting electrons, with repulsive interactions between bosons playing the role of the Pauli exclusion principle. This phase has various unusual properties including a divergent superfluid susceptibility.

The transition from the Bose glass to the superfluid phase will be discussed. Some of the exponents associated with this transition are predicted exactly.

Applications to Helium 4 absorbed in porous media and granular superconductors will be mentioned. The ideas developed are also useful for studying metal insulator transitions in Fermi systems.

[1] D.S. Fisher and M.P.A. Fisher, submitted to *Phys. Rev. Lett.*
[2] M.P.A. Fisher, P. Weichmann, G. Grinstein and D.S. Fisher, preprint.

High–Temperature Superconductivity

L.H. Greene

Bellcore
331 Newman Springs Road
Red Bank, NJ 07701

A revolution in solid–state physics occurred over the past two years. Before March of 1986, superconductivity was only observed in materials at very low temperatures ($T_c \sim 23K$). Superconductivity then was discovered in $La_{2-x}Ba_xCuO_{4-y}$ near 30K, which led to the discovery of $YBa_2Cu_3O_{7-y}$ with $T_c \sim$ 90K. Both these compounds are anisotropic (quasi 2–dimensional) in their structural and transport properties, with metallic–like conduction occurring in parallel CuO_2 planes. The 40K and 90K superconductors have one and two of these layers per unit cell, respectively. Earlier this year, a new series of high–T_c materials was discovered: $Bi_2Sr_2Ca_{n-1}Cu_nO_y$ where n = 1, 2 and 3, the number of CuO_2 planes per unit cell, scales with T_c; 10K, 85K and 110K, respectively. (The Tl-Ba-Ca-Cu-O isomorph shows the highest T_c to date at 125K for $n = 3$.)

This talk will cover experimental results over the last two years, (with a focus on those obtained in our laboratory) that we believe help elucidate the mechanism of high–T_c superconductivity. One problem is that the superconducting coherence length in these materials is surprisingly short ($\sim$ 10 Å) making tunneling measurements extremely difficult. However, a focus upon cation substitutions and oxygen stoichiometry changes have shown that these materials are not conventional BCS–phonon superconductors. First, unlike conventional superconductors, any mechanism proposed must be highly sensitive to local structural order; more local than even the unit–cell size. Unlike conventional superconductors, substitutions for Cu by magnetic ions depress T_c, but no faster than diamagnetic impurity substitutions, and a Curie–Weiss behavior in a material in its superconducting state has been observed. A final interesting point is that parent, non–superconducting, compounds of the La– and Y–based materials exhibit antiferromagnetic order giving some support to a magnetic pairing mechanism.

General References:

[1] *Novel Superconductivity*, S.A. Wolf and V.Z. Kresin, eds., Proc. of the Int'l. Workshop on Novel Mech. of Superconductivity, Berkeley, CA, Plenum Press, New York, 1987.

[2] *High–Temperature Superconductivity*, Proc. Int'l. Conf. on Materials and Mechanisms of Superconductivity, Interlaken, Switzerland, (reprinted from *Physica C* 153–155).

Some References to Bellcore Work:

[1] J.M. Tarascon, L.H. Greene, W.R. McKinnon, G.W. Hull and T.H. Geballe, "Superconductivity at 40K in the Oxygen–Defect Perovskites $La_{2-x}Sr_xCuO_{4-y}$", *Science* 235, 1987, p. 1373.

[2] Y. Le Page, W.R. McKinnon, J.M. Tarascon, L.H. Greene, G.W. Hull and D.M. Hwang, "Room Temperature Structure of the 90K Bulk Superconductor $YBa_2Cu_3O_{8-x}$", *Phys. Rev. B* 34, 1987, p. 7245.

[3] J.M. Tarascon, W.R. McKinnon, L.H. Greene, G.W. Hull and E.M. Vogel, "Oxygen and Rare Earth Doping of the 90K Superconducting Perovskite $YBa_2Cu_3O_{7-x}$", *Phys. Rev. B* 36, 1987, p. 226.

[4] L.H. Greene, J.M. Tarascon, B.G. Bagley, P. Barboux, W.R. McKinnon and G.W. Hull, "Physical Properties of Chemically–Doped High T_c Perovskites", in *Progress in High–Temperature Superconductivity Vol. III*, S.M. Bose and S.D. Tyagi, eds., World Science, Singapore, 1988, p. 53.

[5] P.F. Miceli, J.M. Tarascon, L.H. Greene, P. Barboux, F.J. Rotella and J.D. Jorgensen, "On the role of Bond Lengths in the 90K Superconductor: A Neutron Powder Diffraction Study of $YBa_2Cu_3O_{7-y}$", *Phys. Rev. B* 37,, 1988, p. 5932.

[6] J.M. Tarascon, P. Barboux, P.F. Miceli, L.H. Greene, G.W. Hull, M. Eibschutz and S.A. Shunshine, "Structural and Physical Properties of the Metal (M) Substituted $YBa_2Cu_{3-x}M_xO_{7-y}$ Perovskite", *Phys. Rev. B* 37, 1988, p. 7458.

[7] J.M. Tarascon, Y. LePage, P. Barboux, B.G. Bagley, L.H. Greene, W.R. McKinnon, G.W. Hull, M. Giroud and D.M. Hwang, "Crystal Substructure and Physical Properties of the Superconducting Phase $Bi_4(Sr, Ca)_6Cu_4O_{16+x}$", *Phys. Rev. B* 37, 1988, p. 9382.

[8] J.M. Tarascon, Y. LePage, L.H. Greene, B.G. Bagley, P. Barboux, D.M. Hwang, G.W. Hull, W.R. McKinnon and M. Giroud, "On the Origin of the 110K Su-

perconducting Transition in the Bi-Sr-Ca-Cu-O System", *Phys. Rev. B*, in press.

[9] J.M. Tarascon, W.R. McKinnon, P. Barboux, D.M. Hwang, B.G. Bagley, L.H. Greene, G.W. Hull, Y. LePage, N. Stoffel and M. Giroud, "Preparation, Structure and Properties of the Superconducting Compound Series $Bi_2Sr_2Ca_{n-1}Cu_nO_{8+y}$ with $n = 1$, 2 and 3", *Phys. Rev. B*, in press.

Percolation and Transport in Highly Disordered Systems

B.I. Halperin

Physics Department
Harvard University
Cambridge, MA 02138

Interesting variations of the classical percolation problem occur in systems with a wide range of bond strengths. In collaboration with a student, Stephane Tyc, I have reexamined the problem of the conductivity σ of an infinite resistor network when the conductance of a bond between two neighboring lattice sites is given by $g_{ij} = g_0 \exp(lx_{ij})$, where x_{ij} are independent random variables uniformly distributed on the interval $(-\frac{1}{2}, \frac{1}{2})$, and $l \to \infty$. Define a critical conductance $g_c = g_0 \exp(lx_c)$ such that the conductances with $g_{ij} \geq g_c$ just form a percolating network. We hypothesize that $\sigma \propto g_c l^{-y}$ where the exponent y depends only on the spatial dimensionality d. We have obtained rigorous bounds on σ which imply $(1-s) < y < (t-1)$, where t is the exponent for the vanishing of the conductivity at the percolation threshold of a mixture of unit conductors and insulating bonds, and s is the exponent for the divergence of the conductivity of a mixture of conductors and superconductors. Duality arguments show that the hypothesized form is correct for a square lattice, with $y = 0$, and numerical studies support strongly the idea that $y = 0$ for the triangular lattice. For $d = 3$, the rigorous bounds imply $0.2 \lesssim y \lesssim 0.9$. Estimates of y will be discussed for $d = 3$. I shall also review some other problems with a wide distribution of bond strengths.

[1] S. Tyc and B.I. Halperin, submitted to *Phys. Rev. B Rapid Communications.*

[2] P. LeDoussal, submitted to *Phys. Rev. B Rapid Communications.*

[3] V. Ambegokar, B.I. Halperin and J.S. Langer, *Phys. Rev B* 4, 1971, p. 2612.

[4] M. Pollak and B.I. Halperin, *Solid State Comm.* 13, 1973, p. 869.

[5] V. Ambegaolar, S. Cochran and J. Kurkijarvi, *Phys. Rev. B* 8, 1973, p. 3682.

[6] J. Kurkijarvi, *Phys. Rev. B* 9, 1974, p. 770.

[7] A.L. Efros and B.I. Shklovskii, *Electronic Properties of Doped Semiconductors*, Springer Verlag, Berlin, 1984.

[8] J. Bernasconi, W.R. Schneider, and H.J. Weismann, *Phys. Rev. B* 16, 1977, p. 5250.

[9] S. Feng, B.I. Halperin, and P. Sen, *Phys. Rev. B* 35, 1987, p. 197.

[10] J. Machta, *Phys. Rev. B* 37, 1988, p. 7892.

[11] T.C. Lubensky and A.M.S. Tremblay, *Phys. Rev. B* 37, 1988, p. 7894.

Theory of Martensitic Transformations

R.D. James

Department of Aerospace Engineering and Mechanics
University of Minnesota
Minneapolis, MN 55455

We describe martensitic transformations, noting particularly the fine microstructures observed, and we discuss why they are different from other structural transformations. In some materials the microstructures which appear in the material upon transformation look similar to the laminates used to construct effective materials in homogenization theory. We discuss this similarity, and we describe a theory for martensitic transformations in which minimizing sequences model the fine structure. The microstructure of many martensitic materials is particularly subject to change induced by applying loads or changing the boundary conditions, unlike composite materials. A central and unsolved problem is how these imposed boundary conditions are related to the microstructure.

[1] J.M. Ball and R.D. James, "Fine phase mixtures as minimizers of energy", *Arch Rat. Mech. Anal.* 100, 1987, pp. 13–52.

[2] J.M. Ball and R.D. James, "Experimental tests of a theory of fine microstructures and the two–well problem", preprint.

[3] M.W. Burkart and T.A. Read, "Diffusionless phase change in the indium–thallium system", *Trans. AIME J. of Metals* 197, 1953, pp. 1516–1524.

[4] M. Chipot and D. Kinderlehrer, "Equilibrium configurations of crystals", *Arch. Rat. Mech. Anal.*, to appear.

[5] J.L. Ericksen, "Small martensitic transformations", preprint.

[6] J.W. Gibbs, "On the equilibrium of heterogeneous substances", in *The Scientific Papers of J. Willard Gibbs*, vol. 1, Dover Publications, 1961. (The chapter entitled, "The conditions of internal and external equilibrium for solids in contact with fluids with regard to all possible states of strain of the solids")

[7] R.D. James, "Displacive phase transformations in solids", *J. Mech. Phys. Solids* 34, 1986, pp. 359–394.

[8] R.D. James, "Microstructure and weak convergence", in Proc. of the Symposium on Material Instabilities in Continuum Mechanics, J.M. Ball, ed., Oxford University Press, 1988.

[9] R.V. Kohn and P. Sternberg, "Local minimizers and singular perturbations", *Proc. Roy. Soc. Edinburgh* 111A, 1989, pp. 69–84.

[10] S.W. Meeks and B.A. Auld, "Periodic domain walls and ferroelastic bubbles in neodymium pentaphosphate", *Appl. Phys. Lett.* 47, 1985, pp. 102–104.

[11] L. Tartar, "Étude des oscillations dans les equations aux derivées partielles nonlineaires", in *Lecture Notes in Physics* 195, Springer–Verlag, 1984, pp. 384–412.

[12] J. van Landuyt, G. van Tendeloo, S. Amelinckx, and M.B. Walker, "Interpretation of Dauphiné–twin–domain configurations resulting from the $\alpha-\beta$ phase

transition in quartz and aluminum phosphate", *Phys. Rev. B* 31, 1985, pp. 2986–2992.

[13] M.S. Wechsler, D.S. Lieberman and T.A. Read, "On the theory of the formation of martensite", *Trans. AIME J. Metals* 197, 1953, pp. 1503–1515.

Experiments on the Localization of Light

Ad Lagendijk

Department of Physics
University of Amsterdam
The Netherlands

Multiple scattering is a universal phenomenon occurring in almost any branch of physics. After the discovery of Anderson localization of the electron as an interference effect in multiple elastic scattering in random media the study of multiple scattering of waves in random media has enjoyed a tremendous revival. The ultimate goal is the observation of "localization of light". The idea is to find a medium in which the multiple elastic scattering of light is so gigantic that the mean–free path of light can be made shorter than the wavelength divided by 2π, a criterion known as the Ioffe–Regel rule.

An experimental breakthrough which triggered a lot of activity was the discovery of enhanced backscattering and the recognition of its connection to localization [1,2]. Many of the subsequent activities have been devoted to analyzing these enhanced backscattering effects in great detail [3]. Recently the build–up of interferences in multiple scattering has been observed in the time domain with femto–second time resolution. The connection between weak localization and strong localization is that enlarged backscattering of a diffusing wave packet leads to an effective reduction in the diffusion constant of the wave packet. When the influence of the increased backscattering is so overwhelming that the diffusion constant vanishes, (strong) localization sets in.

Unfortunately there are several severe complications hampering its observation. It turns out that it is only in very dense samples with a large contrast in index of refraction fluctuating on a length scale of the order of the wavelength that one can approach the Ioffe–Regel boundary.

As dimensionality is an essential parameter and localization is easier to induce in lower dimensionality much effort is devoted to the development of two–dimensional specimens. Many loss mechanisms have to be overcome before critical localization effects can be observed in these materials.

The analogy between (quantum) electrons and (classical) waves runs indeed very deep. The most striking similarity is beyond any doubt the relation between universal conductance fluctuations and optical speckles [4]. When the disorder is static so that an experimentalist essentially measures only one "realization" of the disorder in a scattering experiment huge fluctuations are present. Many workers are investigating statistical properties of optical speckles and are examining their universal behavior.

A book containing contributions of leading workers in this field will be published shortly providing a good introduction to this rapidly developing area of research [5].

[1] M.P. van Albada and A. Lagendijk, *Phys. Rev. Lett.* 55, 1985, p. 2692.
[2] P.E. Wolf and G. Maret, *Phys. Rev. Lett.* 55, 1985, p. 2696.
[3] See for instance: M.B. van der Mark, M.P. van Albada, and A. Lagendijk, *Phys. Rev. B* 37, 1988, p. 3575.
[4] S. Feng et al., *Phys. Rev. Lett.* 61, 1988, p. 834; M. Stephen and G. Cwillich, to appear.
[5] *Classical Localization*, P. Sheng, ed., World Scientific, Singapore, to appear.

Localization of the Sound Modes of Helium in Porous Media and Symmetry Breaking by a Flow

J. Machta

Department of Physics and Astronomy
University of Massachusetts
Amherst, MA 01003

The localization of acoustic waves by static elastic scatterers is reviewed [1] and experiments on third and fourth sound in superfluid Helium in porous media are discussed [2, 3]. The effect of adding a steady, uniform flow field is examined [4]. A flow field breaks time reversal invariance and leads to a modification of critical exponents for $d \geq 2$. For example, for $d = 2$ all modes are localized with or without a flow but the localization length as a function of frequency diverges like $\exp(1/f^4)$ with a flow in contrast to $\exp(1/f^2)$ without a flow. This effect should be visible in third sound experiments on disordered substrates.

These results are obtained using the field theoretic formulation of the localization transition wherein the flow causes a crossover from the orthogonal to the unitary fixed point. The effect of a flow in acoustic localization is analogous to the effect of a magnetic field in electronic localization; however the flow case leads to a simpler field theory.

Helium in porous media promises to be a useful system in which to study localization and the effects of symmetry breaking fields. A flow field provides a new experimental tool in studying classical wave localization and in probing fundamental questions in the theory of the Anderson transition.

This work was done in collaboration with S.M. Cohen and T.R. Kirkpatrick, and was supported in part by the National Science Foundation.

[1] S. John, H. Sompolinsky and M.J. Stephen, *Phys. Rev. B* 27, 1983, p. 5592; T. R. Kirkpatrick, *Phys. Rev. B* 31, 1985, p. 5746.
[2] S.M. Cohen and J. Machta, *Phys. Rev. Lett.* 54, 1985, p. 2242.
[3] D.T. Smith, C.P. Lorenson, R.B. Hallock, K.R. McCall, and R.A. Guyer, "Third sound on substrates patterned with periodic and random disorder: Evidence for classical wave localization", *Phys. Rev. Lett.*, to appear.

[4] S.M. Cohen, J. Machta, T.R. Kirkpatrick and C.A. Condat, *Phys. Rev. Lett.* 58, 1987, p. 785.

Acoustic Anderson Localization

J.D. Maynard

Department of Physics
The Pennsylvania State University
University Park, PA 16802

As a consequence of Bloch's theorem, the eigenstates of a system described by a periodic potential are extended, having the same nominal amplitude at all positions in space. When a random element is added to the potential field, the eigenstate amplitude may not simply acquire a random variation but instead may become localized, decaying exponentially from a particular site with a characteristic length. This phenomenon was originally discussed by Anderson [1] and Mott [2] in reference to electron Schrödinger waves in a disordered solid. The Anderson localization effects are strongest in one–dimensional systems, and there has been considerable experimental effort in condensed matter physics devoted to observing clear effects for electrons in narrow wires and semiconductor channels [3]. However the measurements are difficult, and the quantum waves cannot be measured directly. It is possible to observe localization effects in other wave systems, such as ones involving electro–magnetic waves [4] or phonons [5] where observations may be more straightforward. We have been making such measurements in an acoustic localization experiment where we can observe in detail eigenstate amplitude and phase, eigenvalue spectra and density of states, and dependence of localization length on strength of disorder, showing fluctuations [6] as a function of position within a band, separation from nearby states, etc.

In the acoustic experiment the wave medium was a long ($\sim$ 15 m) steel wire suspended vertically with a tension maintained with a weight at the bottom. The wave field analogous to the electron Schrödinger waves consisted of transverse waves in the wire generated with an electromechanical actuator. The random potential field (analogous to disordered ion sites in the mesoscopic wire) was provided by small lead masses spaced along the wire. When the lead masses were placed as periodic positions, Bloch wave eigenvalues and extended eigenfunctions were verified; when the positions were disordered, the eigenvalue spectrum changed dramatically, and Anderson localized states were observed. The presence of a lattice vibration (phonon) was simulated by using a second electromechanical actuator to modulate the tension in the wire; the resulting strain modulation causes the mass lattice sites to oscillate about their equilibrium positions, analogous to a longitudinal phonon in a solid. When the phonon frequency (the drive frequency of the second actuator) matched the difference frequency between two of the Anderson states localized at different sites, resonant hopping between the two sites could be observed, and the hopping probability was measured [7].

[1] P.W. Anderson, *Phys. Rev.* 109, 1958, p. 1492.
[2] N.F. Mott, *Adv. Phys.* 16, 1967, p. 49.
[3] A.B. Fowler, G.L. Timp, J.J. Wainer, and R.A. Webb, *Phys. Rev. Lett.* 57, 1986, p. 138.
[4] M.J. Stephen, *Phys. Rev. Lett.* 56, 1986, p. 1809.
[5] S. John, H. Sompolinsky and M.J. Stephen, *Phys. Rev. B* 27, 1983, p. 5592; M.L. Williams and H.J. Maris, *Phys. Rev. B* 31, 1985, p. 4508.
[6] P.A. Lee and A.D. Stone, *Phys. Rev. Lett.* 55, 1985, p. 1622.
[7] Shanjin He and J.D. Maynard, *Phys. Rev. Lett.* 57, 1986, p. 3171.

Synkinematic Microscopy of Transparent Polycrystals

W.D. Means

Department of Geological Sciences
State University of New York at Albany
1400 Washington Avenue
Albany, NY 12222

Deformation of weak, polycrystalline materials in thin section, on the stage of a polarizing microscope, offers insight into the origin of optical microstructures similar to microstructures in naturally deformed rocks and other crystalline materials. Sample materials used to date include several soft minerals (ice, carnallite, bischofite, nitratite) and various organic plastic materials. They can be simple–sheared or pure–sheared at temperatures to 300°C, at a variety of strain rates, using relatively simple and inexpensive apparatus. The complete microstructural evolution can be recorded photographically, along with the complete kinematic history. Research applications have focussed on dynamic recrystallization, and progress has been made toward understanding how the average grain size is maintained constant in steady–state recrystallization flow. Intimate views have also been obtained of the development of various crystal–plastic microstructures and of lattice reorientation trajectories during flow. Teaching applications have begun to be developed, and several videotapes are available. Future technical developments are expected to permit stress measurement or control during experiments and synkinematic recording of lattice reorientation. Numerous research areas invite attention. These include investigation of the microstructural developments that precede shear fracture, the behavior of deforming crystal/melt systems, and the behavior of dynamically reactive, multi–phase systems.

[1] M.W. Jessell, "Grain boundary migration and fabric development in experimentally deformed octachloropropane," *J. Struct. Geol.* 8, 1986, pp. 527–542.
[2] W.D. Means, "Microstructure and micromotion in recrystallization flow of octachloropropane: A first look", *Geol. Rdsch.* 72, 1983, pp. 511–528.
[3] W.D. Means and M.W. Jessel, "Accomodation migration of grain boundaries", *Tectonophysics* 127, 1986, pp. 67–86.

[4] W.D. Means and Z.G. Xia, "Deformation of crystalline materials in thin section", *Geology* 9, 1981, pp. 538–543.
[5] P.D. Tungatt and F.J. Humphreys, "Transparent analogue materials as an aid to understanding high temperature polycrystalline plasticity", in *Deformation of Polycrystals: Mechanisms and Microstructures*, A. Horsewell, T. Leffers and H. Lilholt, eds., Riso Nat. Lab., Roskilde, 1981, pp. 393–398.
[6] P.D. Tungatt and F.J. Humphreys, "The plastic deformation and dynamic recrystallization of polycrystalline sodium nitrate", *Acta Metall.* 32, 1984, pp. 1625–1635.
[7] J.L. Urai, "Development of microstructure during deformation of carnallite and bischofite in transmitted light", *Technophysics* 135, 1987, pp. 251–263.

Infrared Electrodynamic Properties of High T_c Superconductors

T.W. Noh

Laboratory of Atomic and Solid State Physics and Materials Science Center
Cornell University
Ithaca, NY 14853

The IR spectra of oxide superconductors in sintered polycrystalline, thin film, and single crystal form have continued to evolve over the last year as sample preparation techniques have become better defined. Over this same time period there has been increasing interest in such IR measurements because of the realization that the small superconducting coherence length in these materials makes the surface sensitive electron tunneling spectroscopic method an unlikely probe with which to monitor bulk properties. Therefore measurement of the infrared properties of the new high T_c superconductors is the direct method of characterizing the electrodynamic properties in the normal state carriers and of identifying the energy gap and possible coupling mechanisms in the superconducting state. The technique consists of applying optical and IR Fourier transform spectroscopic techniques to reflectivity and direct absorption measurements over an extended frequency interval. Kramers–Kronig and mode analysis of these data give the real and imaginary parts of the frequency dependent conductivity, a quantity which can be compared directly with theoretical predictions. A particularly novel and interesting electrodynamic study has been that associated with the analysis of the data on sintered pellets composed of randomly oriented highly anisotropic grains. The effective medium approximation (EMA) with an anisotropic dielectric function has been developed and successfully used to explain the far–infrared response of the sintered high T_c superconductors. It is shown that the EMA applied to the nonspherical grains with Lorenz oscillator contributions for the phonons and a two–dimensional Drude free–carrier conductivity provides a good description of the far–infrared properties of these materials. By modeling the effective conductivity in the superconducting state with the Mattis–Bardeen equations in the EMA, we can explain the temperature dependence of the measured reflectivity. Moreover, we find a strong dependence on the grain geometry of both the apparent gap frequency and the magnitude of the excess absorptivity in the superconducting state above the gap. The analysis of the superconducting state data in the sintered polycrystalline, thin film and single crystal form shows that absorption occurs at frequencies below the apparent electromagnetic "energy gap" and that the effective gap position depends on the sample texture.

This work was done in collaboration with S.G. Kaplan, P.E. Sulewski and A.J. Sievers.

[1] P.E. Sulewski, T.W. Noh, J.T. McWhirter, and A.J. Sievers, *Phys. Rev. B* 36, 1987, p. 5735.
[2] T.W. Noh, P.E. Sulewski, and A.J. Sievers, *Phys. Rev. B* 36, 1987, p. 36.
[3] T.W. Noh, S.G. Kaplan, H. Xia, and A.J. Sievers, to appear.

Localization of Classical Waves in a Random Two Component Medium

C.M. Soukoulis

Ames Laboratory–USDOE
Department of Physics
Iowa State University
Ames, Iowa 50011

Various calculational methods for studying the problem of electronic localization will be presented. An exact mapping of the propagation of classical waves in random media to that of electrons in random systems is developed [1]. Extensive numerical results are used to examine the existence of genuine localization (i.e., one occurring when both components have real dielectric constants) of classical waves in random binary alloys $A_{1-x}B_x$. It is found that scalar waves do exhibit localization [1]. A Coherent Potential Approximation (CPA) is also developed which for $x < 0.2$ gives results very similar to the numerical ones. This CPA can be easily generalized to electromagnetic fields as well, where explicit results for transport properties will be presented [1,2]. These CPA results [3] are in surprising impressive agreement with recent experimental data [4] close to the critical region and indicate ways of experimentally obtaining classical localization.

[1] C.M. Soukoulis, E.N. Economou, G.S. Grest and M.H. Cohen, *Phys. Rev. Lett.* 62, 1989, p. 575.
[2] E.N. Economou and C.M. Soukoulis, "Scattering and Localization of Classical Waves in Random Media", in *Classical Localization*, Ping Sheng, ed., World Science, 1989.
[3] E.N. Economou and C.M. Soukoulis, *Phys. Rev. B*, to appear.
[4] J.M. Drake and A.Z. Genack, *Phys. Rev. Lett.* 63, 1989, p. 259.

Theory of the Transport and A.C. Properties of Superconducting Composites

D. Stroud

Department of Physics
Ohio State University
Columbus, Ohio 43210

The theory of d.c. and a.c. transport in superconducting composites will be reviewed, particularly within the context of the effective medium approximation. We will discuss percolation in a superconducting–normal (S/N) composite, and the conductivity and Hall coefficient of a highly anisotropic superconducting composite, such as some of the new high–temperature superconductors. In the finite–frequency range, we will discuss the theory of the optical properties and infrared properties of uniaxial optical materials prepared in polycrystalline form, and we will discuss the application of these results to anisotropic superconductors. We will also discuss the behavior of the superfluid density of percolating superconductors, and its relation to the penetration depth of inhomogeneous superconductors.

Infrared Studies of High-T_c Materials

D.B. Tanner

Department of Physics
University of Florida
Gainesville, FL 32611

Measurements of the infrared properties of the high–T_c superconductors $YBa_2Cu_3O_{7-\delta}$ and $La_{2-x}Sr_xCuO_4$ have been carried out between 300K and 4K. Both oriented samples with polarizations parallel to the two–dimensional conducting Cu-O layers (e.g., textured ceramics, oriented films, and small crystals) and random ceramic samples have been studied. In both cases, a strong, temperature–independent absorption is observed in the mid–infrared (0.1 to 0.5 eV or 800 to 4000 cm^{-1}) energy range, with a temperature–dependent Drude contribution in the far infrared (below 0.1 eV/800 cm^{-1}). The strength of the mid–infrared absorption is correlated with the superconductivity whereas the Drude portion accounts for the dc conductivity.

In the superconducting state a gap–like depression occurs in the far–infrared Drude conductivity, but the gap is not complete, i.e., there remains a finite absorption even at the lowest temperatures. The reflectance in the superconducting state shows a significant decrease at 500 cm^{-1}; this, however, is associated with dispersive response of the superconducting electrons rather than with any energy gap structure. The reflectance drop is interpreted as a "plasma edge" of the superconducting carriers, sufficiently screened by the positive contribution to the dielectric constant of the mid–infrared absorption that it appears in the far infrared rather than at the ~ 1 eV bulk plasma frequency.

This work was done in collaboration with S.L. Herr, K. Kamaras, and C.D. Porter (Univ. of Florida, Gainesville); D.A. Bonn, A.H. O'Reilly, J.E. Greedan, C.V. Stager, and T. Timusk (McMaster University); and S. Etemad and Siu–Wai Chan (Bell Communication Research). Supported by DARPA grant MDA872–88–J–1006, by the NSF–Solid State Chemistry–DMR 8416511, and by NSERC of Canada.

Pulse Reflection from a Randomly–Stratified Medium

Benjamin S. White

Applied Math Group
Exxon Research and Engineering LD 352
Rt. 22E, Annandale, NJ 08801

We compute the statistics of the random signal generated when an acoustic pulse is backscattered from a randomly inhomogeneous plane–layered medium. It is assumed that the medium is composed of processes with two spatial scales: a fine–scale stochastic microstructure with statistics which vary slowly, i.e. with trends which vary on a much longer scale than the random fluctuations. It is not assumed that the magnitude of the random fluctuations is small. The medium is probed with a pulse whose spatial extent is intermediate between the two spatial scales of the medium. Since the pulse is too broad to reliably detect the stochastic microstructure, the backscattered field consists of a "signal" produced by the smooth trends, which is corrupted by the "noise" of random multiple scattering. We thus have a description of the backscatter in which signal and noise are modelled from first principles.

Using limit theorems for stochastic differential equations with multiple time scales, we derive approximate statistics for the above process. It is shown that when the time series is segmented into "time windows" of appropriate size, the resulting process segments are stationary Gaussian processes with power spectra which can be computed from averages of the medium statistics. These power spectra vary among windows on a time scale comparable to traverse times of the medium trends. Explicit results are computed for homogeneous media, where connection can be made to localization theory.

Transport Properties of Real and Artificial Rocks

Po–zen Wong

Department of Physics and Astronomy
University of Massachusetts
Amherst, MA 01003

Sedimentary rocks are generally made up of cemented sand grains or carbonate grains. The random packing of these grains creates a network of connected pores which are filled with conductive water and insulating hydrocarbons. Among the most important problems in the oil industry are the determination of rock porosity (θ), the water saturation (S_w), the oil saturation (S_o), and the rock permeability (k). The most common techniques for estimating these quantities involve the use of conductivity (σ) and dielectric (ϵ) measurements. This paper presents laboratory data of such measurements on a large number of real and artificial rocks over a wide range of porosities with $S_w = 1$. Such systems can be considered as random two–component composites. It is shown that much of their transport properties can be explained by effective medium theories [1] and simple network models [2]. However, a major shortcoming exists for shaly sandstones in which the presence of clay particles creates a space charge layer at the pore–grain interface [3]. Both the d.c. conductivity and a.c. dispersions exhibit scaling behavior which may be related to the fractal character of the interface [4]. The importance of such interfacial phenomena in composites is emphasized.

[1] P.N. Sen, C. Scala, and M.H. Cohen, *Geophysics* 46, 1981, p. 781; R. Landauer, in *Electrical Transport and Optical Properties of Inhomogeneous Media*, AIP Conf. Proc. 40, J.C. Garland and D.B. Tanner, eds., AIP, New York, 1978.

[2] See e.g., P.–z. Wong, *Physics Today*, to appear, and references therein.

[3] W.C. Chew and P.N. Sen, *J. Chem. Phys.* 77, 1982, p. 4683; D.L. Johnson and P.N. Sen, *Phys. Rev. B* 37, 1988, p. 3502.

[4] See e.g., P.–z. Wong, in *Physics and Chemistry of Porous Media II*, AIP Conf. Proc. 154, J. Banavar, J. Koplik and K. Winkler, eds., AIP, New York, 1987, and references therein.

SIAM WORKSHOP ON RANDOM MEDIA AND COMPOSITES

December 7–10, 1988

Xerox Training Center, Leesburg, VA

WEDNESDAY, DECEMBER 7

Session 1: PERCOLATION

Percolation and Transport in Highly Disordered Systems, B. Halperin, Harvard University

Simulations of Composites: From Ohmic Response to the Breakdown of Superconductivity, C. Lobb, Harvard University

Convexity in Random Resistor Networks, K. Golden, Princeton University

Nonlinear Behavior and 1/f Noise Near a Conductivity Threshold—The Importance of Local Geometry and the Failure of Effective Medium Theory, D. Bergman, Tel–Aviv University

Session 2: EFFECTIVE MODULI I

An Effective Continuum Theory for Estimating Transport Parameters in Two–Phase Random Media, A. Acrivos, City University of New York

An Intuitive Approach and a Diagrammatic Formulation of the Renormalized Polarizability in the Maxwell–Garnett Theory, R. Barrera, Universidad Nacional de Mexico

Estimating Effective Moduli of Composites Using Quantitative Image Analysis, J. Berryman, Lawrence Livermore National Laboratory

Transport Properties of Real and Artificial Rocks, P.Z. Wong, University of Massachusetts

Critical Behavior of Effective Moduli Determined from Nearest Neighbour Interactions, L. Poladian, University of Sydney

THURSDAY, DECEMBER 8

Session 3: EFFECTIVE MODULI II

Bounds on Various Electrostatic and Hydrodynamic Capacities, J. Rubinstein, Israel Institute of Technology

Transport and Dispersion in Random Media, S. Redner, Boston University

Optimal Bounds and Microgeometries for Polycrystalline Aggregates, Marco Avellaneda, New York University

Homogenization in Thermoelasticity, F. Murat, University of Paris VI

H-Measures and Small Amplitude Homogenization, L. Tartar, Carnegie-Mellon University

Session 4: NONLINEAR EFFECTS

Theory of Martensitic Transformations, R. James, University of Minnesota, Minneapolis

Synkinematic Microscopy of Transparent Polycrystals, W. Means, State University of New York

The Overall Behaviour of a Nonlinear Matrix Reinforced by Rigid Inclusions, or Weakened by Cavities, John Willis, University of Bath

Scaling Theory and Fluctuations in the Strength of Composites, P.M. Duxbury, Michigan State University

FRIDAY, DECEMBER 9

Session 5: SUPERCONDUCTIVITY

Infrared Electrodynamic Properties of High T_c Superconductors, T.W. Noh, Cornell University

Theory of the Transport and A.C. Properties of Superconducting Composites, D. Stroud, Ohio State University

High Temperature Superconductivity, L. Greene, Bell Communications Research

Infrared Studies of High-T_c Materials, D. Tanner, University of Florida

Session 6: LOCALIZATION I

Pulse Reflection from a Randomly-Stratified Medium, B. White, Exxon Research and Engineering Co.

Localization of Classical Waves in a Random Two-Component Medium, C.M. Soukoulis, Iowa State University

Onset of Superfluidity in Random Media, D. Fisher, Princeton University

Propagation of Regularity and Inverse Problems, W. Symes, Rice University

SATURDAY, DECEMBER 10

Session 7: LOCALIZATION II

Experiments on the Localization of Light, A. Lagendijk, University of Amsterdam

Acoustic Anderson Localization, J. Maynard, Pennsylvania State University

Localization of the Sound Modes of Helium in Porous Media and Symmetry, Breaking by a Flow, J. Machta, University of Massachusetts